What is Empty Space?

Some Current Mysteries in Physics and Cosmology

A short, easy read for the non-physicist

By
Doug Domke

Contents

Preface

In 2016 physics and cosmology are in the same condition as they were in back in 1885. At that time, there were many problems associated with the nature and speed of light. The more experimental data that was collected, the more confusing the situation became. It wasn't until twenty years later, when Albert Einstein came along, that all these mysteries started to be resolved with the introduction of special relativity.

Today, we have our own set of mysteries and seemingly conflicting results that no one fully understands. We will probably need another great mind, like an Einstein, and another big breakthrough in the next few years to make sense of it all.

Why ask what empty space really is? It is the nature of space itself that seems to be at the heart of the mysteries in physics today.

Let's think about the nature of space. We know that the Earth is moving because we are in orbit around the Sun. But we don't really have a sense of that motion, except by seeing the evening stars change position in the sky over the course of a year. Motion is relative. You can't tell you are moving except in relation to other objects.

But what about spinning? When a skater does a spin, she feels her arms pulled outward by centrifugal force. That outward pull is just the result of spinning. It's not relative to anything. Picture the skater in the empty void of space with nothing around to tell her she is spinning. Yet she would still feel the pull on her arms if she was spinning. What is she spinning relative to? Empty space is the answer, so it's somehow more than the absence of anything!

Einstein also told us that space is a something. He showed it could be stretched and warped when he published general relativity in 1915. But over the last one hundred years, the nature of empty space has continued to become more mysterious and confusing. For example, we have found out that the expansion of the universe is accelerating for reasons we don't understand. We say this acceleration is powered by a mysterious force called dark energy, which seems to be a characteristic of empty space.

Quantum theory tells us that empty space contains particles that momentarily burst into existence and then disappear. In recent years we have even been able to measure the minute pressure that these particles can exert on real everyday matter.

Quantum theory, which has been validated by all kinds of experimental evidence, also seems to conflict with

Einstein theories of relativity in many ways. "All our experiences tell us we shouldn't have two dramatically different conceptions of reality — there must be one huge overarching theory," says Abhay Ashtekar, a physicist at Pennsylvania State University. This, in itself, is one of physics great mysteries, and at the heart of it, is our lack of understanding of empty space.

In this book, we will talk about these mysteries, as well as others. These mysteries are, in fact, the current leading edge of physics today. We will discuss what physicists know and what they still don't know. We will attempt to speculate on what it all means. We will discuss a series of observations made during the last fifty years. Some conflict with what we think we know. And some of these observations conflict with each other. We will try to put into context what this all means.

We will also try to put all this in historical context, i.e. what have we learned in the last hundred years and how we learned it. We'll explain how Einstein changed our perception of reality and how that relates to the mysteries we face today.

This book is written for someone who wants to learn a little about physics, but isn't a physicist. That means we will try to explain all of this without delving Into all the mathematics which is at the heart of how physics attempts to describe nature and reality.

Until the last few hundred years, we didn't even have a concept of empty space. According to Aristotle, the Greek philosopher and scientist, the universe consisted of earth, wind, water and fire! We didn't know that the space between the planets and stars was empty, let alone consider what exactly is implied by "empty space". Only recently, did we realize that empty space existed. And now we are coming to realize, we don't really know what it actually is!

"The vacuum is one of the places where our knowledge fizzles out and we're left with all sorts of crazy-sounding ideas," says John Baez, a mathematical physicist at the University of California at Riverside.

That is the subject of this book.

Doug Domke
May 2016

Acknowledgements: Special thanks to my wife Polly and our daughter Ginger Sainz for their suggestions and help with editing.

Chapter 1– The Aether

To understand where we are in physics today, we need to start with some history – what we have learned in the past few hundred years. Once that is clear, we can talk about what we still don't know. In this chapter, we are going to explore what we learned prior to Einstein and his theory of relativity. It is mainly the story of what we learned about light that lead us to Einstein's great discoveries.

Before about 1500, very little was actually known about the universe. We didn't know that the space between the planets and stars was empty. We didn't know why the objects in the sky moved in the way they did. The heavens were a complete mystery.

As people like Nicolaus Copernicus (1473-1543) and Johannes Kepler (1571-1630) studied the heavens, some of the motions of the moon and planets became clearer. It started to become obvious that the Earth was not at the center of everything we could see in the heavens. But what was really going on remained a mystery for the most part.

But all that changed with Sir Isaac Newton! Newton was a brilliant British philosopher and scientist who lived from 1643 until 1727. He was truly the father of modern physics. The list of his accomplishments could fill an entire book. He developed the basic laws of motion, and invented calculus as the mathematical framework with which to express those laws.

Newton developed an equation for gravity, which finally explained the motion of heavenly bodies. His work validated the theory that planets, including the Earth, rotate in orbits about the Sun, and that they do so in empty space, where there is no friction to slow them down or make their orbits decay.

Isaac Newton

Newton did much, much more, but a couple of things are worthy of particular mention here. He treated space as a fixed framework. It was the static background in which objects moved. It's the way we all think of space for the most part. But as we will talk about in the next chapter, it's a very different view than the one that Einstein came up with 250 years later. Einstein saw space as dynamic – much more than a static background.

Newton also did a lot of work with light. He was the first to realize the white light was made up of a combination of colored light, and he made numerous contributions in the field of optics, using lenses and prisms to manipulate light. He also suggested that light had particle-like behaviors, based on how it bounced off mirrors.

But neither Newton nor the scientists who followed him in the next one hundred and fifty years knew anything about what light actually is or why it behaves like it does.

All that changed with James Maxwell, a Scottish physicist who lived from 1831 until 1879. Maxwell was the first to recognize that electricity and magnetism were both a single phenomenon. In 1865, he derived a set of equations that define

James Maxwell

how the two are related and interact with one another. He demonstrated how magnetic fields generate electrical fields, and how electrical fields generate magnetic fields. He showed that a magnetic field and electric field together could propagate as a wave – an electromagnetic wave, and calculated that these waves should travel at a specific speed. That speed had already been established as the speed of light, so Maxwell realized that light itself must be a form of these electromagnetic waves.

Maxwell realized one other thing - the speed of light comes out of his equations without any frame of reference. That seemed odd to Maxwell. For example, if I am walking along at 2 miles per hour, that speed is relative to the surface of the Earth. If I viewed my speed from the Sun, I would be moving along at 18 miles per second – the speed at which the Earth travels as it orbits the Sun. So Maxwell himself was the first to realize that something was very strange about the speed of light.

Scientists in the 19th century knew a fair amount about waves. They knew that waves could propagate though water, just as they do in the oceans. And they knew that sound is waves that propagate through air. They reasoned that these new electromagnet waves must have some medium in which to travel through, just like ocean waves and sound. They also knew that light travels vast distances through "empty space", so they concluded that all of space must be filled with something that light waves travel on. They called it "the aether" (pronounced ee-ther). The aether had to permeate all of space, but also had to allow planets and other heavenly bodies to move freely through it without friction. So empty space wasn't really empty! The aether also seemed to solve Maxwell's mystery of what the speed of light was relevant to.

By 1875, most scientists were convinced that the aether was real, but its existence was still unproven. They knew that if light travelled through the aether, that the speed of light must be relative to the aether, not relative to the Earth that is itself moving through

Albert Michelson   Edward Morley

the aether as we orbit the Sun. If this were true, they should be able to detect small variations in the speed of light, depending on whether they measured it in the direction of our motion through the aether or say at 90 degrees to that motion. What was needed to conduct this experiment was a really accurate way to measure differences in the speed of light.

The now famous Michelson–Morley experiment was performed over the spring and summer of 1887 by

Americans Albert A. Michelson and Edward W. Morley. It compared the speed of light in perpendicular directions, in an attempt to detect the Earth's relative motion through the aether. The Earth's motion through the aether should result in a perceived aether wind that would cause a shift in the speed of light.

Although the equipment was easily sensitive and accurate enough to detect the expected difference in the speed of light, none was detected. Light travelled at exactly the same speed in all directions. Something was very wrong with the aether theory! The experiment was repeated numerous times by numerous people, always with the same result! These results said aether was not the answer. The aether did not exist!

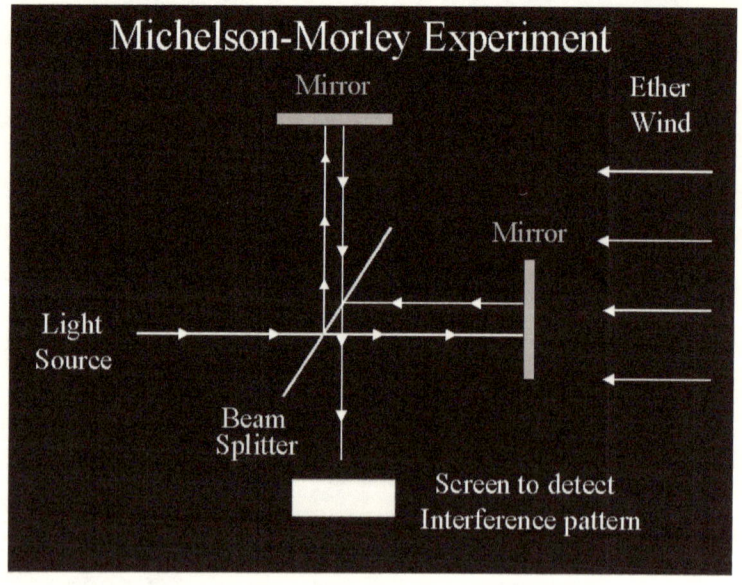

Over the next few years, now that the idea of aether had been discredited, scientists attempted to come up with other theories. One, called the emission theory, suggested that the speed of light was relative to the source of its emission, but independent of any medium through which it travels. It's a simple enough idea, and it was consistent with Newton's basic equations of motion. As an analogy, suppose a gun shoots a bullet at 1000 feet per second, but it's mounted on a truck moving down the road at 100 feet/sec. Shoot the gun in the forward direction of the truck and the bullet travels at 1100 feet/sec. relative to the ground. Shoot the gun out the back and the bullet's speed is 900 feet/sec. relative to the ground.

But as our experimental measurements of light's speed continued to get more and more accurate, it became more and more obvious that the speed of light never changed, no matter where it came from or how it was measured.

As the 19th century came to a close, physics had a real mystery on its hands. It seemed impossible for the speed of light to always be invariant. It seemed to say that the speed of light as we measure it is always relative to us, regardless of where it came from, and regardless of the fact that we are on planet revolving around the sun and also rotating about itself. It just didn't make any sense. It seemed to conflict with the basic foundations that physics had always taken for granted – namely that space and time were universal and not subject to change or interpretation.

Chapter 2 – Special Relativity

So this was the situation in physics when Albert Einstein came upon the scene.   He was actually one of many people trying to make sense of the light speed dilemma.

Einstein was born in Germany in 1879.   He attended the Swiss Federal Polytechnic Institute in Zurich, Switzerland, training to be a teacher of physics and mathematics. He became a Swiss citizen in 1901. Unable to find work as a teacher, however, he moved to Bern and took a  job in the Swiss Patent Office.

In 1905, while still a relative unknown in the world of physics, Einstein published a series of papers and quickly made a name for himself. One described and explained the photoelectric effect, where light hitting certain materials can produce an electric current. Ironically, when Einstein finally won a Nobel Prize in 1921, it was for this work on the photoelectric effect, and not for relativity, which by then had made him the most famous scientist in the world!

In another 1905 paper, in what was actually an outgrowth of his work on the photoelectric effect, Einstein explained the relationship between mass and energy, introducing his now famous equation $E = mc^2$, where E is energy, m is mass, and c is the speed of light. By any normal unit of measure, the speed of light squared is a really big number, so his equation meant that a small amount of mass is equivalent to a very large amount of energy.

But the biggest accomplishment of 1905 was his paper on special relativity where he introduced the idea that space and time are not absolutes, but change relative to motion. And with this comes an explanation for why the speed of light will always be measured the same.

We will spend a little time here discussing special relativity. It is the first time we really see space as something special rather than the mere absence of anything. Einstein describes time and space together as a four dimensional construct he called space-time. It could be stretched, warped, or compacted by motion and matter. Special relativity dealt specifically with how space-time is affected by motion. The distortion of space-time by matter will be the topic of general relativity discussed in the next chapter.

For special relativity, Einstein started out with a simple premise. All the things we have observed must be true, even if they appear to contradict one another. Therefore, in the absence of motion or with motion which is slow compared to the speed of light, space and time both appear normal, as we expect them to be. But at or near the speed of light, both are very different, specifically allowing the speed of light to always be measured as 186,000 miles per second, regardless of one's speed or frame of reference.

What does that mean? Very strange things! Suppose space is constant, and I am travelling away from the sun at 90% of the speed of light. I look out the window at the light streaming away from the sun. I still measure its speed as 186,000 miles per second. For that to be true, time itself must have slowed down for me by 90%. Because otherwise I would be travelling next to the light at 90% of its speed and perceive the light's relative speed as only 10% of what I did back on Earth. So one immediately comes to the realization that time and/or space cannot be the constants we have always assumed them to be.

In the paragraph above, we made the assumption that space was a constant, and that time slowed down by

90% to make the speed of light a constant.   We could also have assumed that time was constant, and if so, space itself must have contracted by 90% for me to measure the speed of light at 186,000 miles per second.

Both the examples above are equivalent – space-time itself has changed as a result of my change in frame of reference.   But that is true only when my measure of time and distance are observed from an external frame of reference.   To me, in my own frame of reference, everything just looks normal.   So time and space are only relative to a particular frame of reference – hence the name of Einstein's theory – relativity.

Einstein was able to develop a set of fairly simple equations that showed how his assumptions worked. At speeds well below the speed of light, space and time do appear more or less constant and absolute. As speeds become a significant fraction of the speed of light, however, that is no longer the case.

Here are some of Einstein's conclusions:

- When viewed from another frame of reference, time slows as speed increases.

- When viewed from another frame of reference, objects shorten as speed increases. Space itself contracts.

- More and more energy is required as an object approaches the speed of light - it takes infinite energy to reach the speed, as though the mass of the object becomes infinite at the speed of light.

- Nothing can travel faster than the speed of light.

- Time travel into the future is possible if you can travel at high enough speeds.

In 1905, physicists around the world were skeptical. However, a wide range of experimental evidence has gradually and overwhelmingly proved that Einstein was correct!

Chapter 3 - General Relativity

Einstein published a number of papers between 1905 and 1915 while he attempted to integrate the effects of gravity and acceleration into his special relativity theory. This effort culminated in 1915 with his theory of general relativity and the Einstein field equations that described the relationships between matter, time, and space. Before going into general relativity in detail, let's take a brief look at how he got there.

In 1907, Einstein published a paper where he described the "equivalence principle". He said that the force we feel from gravity is, in every way, indistinguishable from the force felt when accelerating.

He then began to develop the implications of this equivalence. If gravity is really equivalent to acceleration, and if motion affects measurements of time and space (as shown in special relativity), then it follows that gravity does so as well. In particular, the gravity of any mass, such as our sun, has the effect of making space shorter (warping it) and making clocks run slower.

This view of gravity was very different than that of Isaac Newton. Newton viewed gravity as simply a force in nature that made objects with mass attract each other. Einstein ascribed the effects of gravity to the curvature of space-time instead of a force. Einstein proposed that space-time is curved by matter, and that free-falling objects are moving along locally straight paths in curved space-time.

This concept is a little hard to grasp; a picture helps a lot:

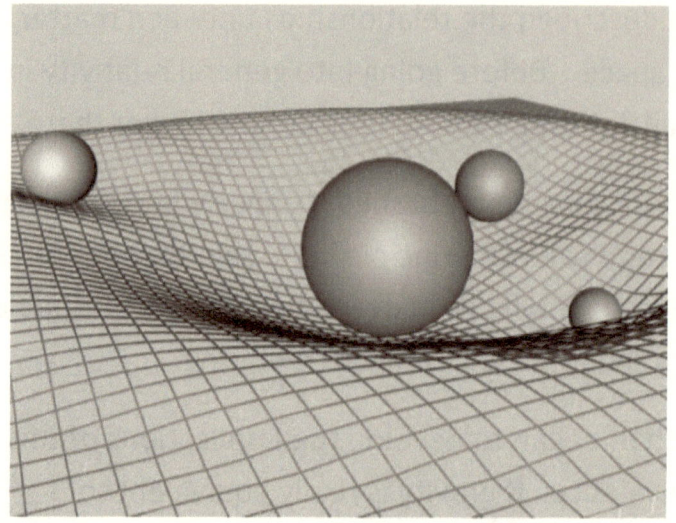

Again, Einstein says that the objects are just trying to follow straight lines, but that the space itself is warped by matter, causing objects to move around each other in a set of behaviors we know as gravity.

Again, in our everyday lives, Newton's version of gravity works perfectly, even when applied to very large objects such as galaxies or very small objects like an atom. But under extreme conditions like approaching the speed of light or extremely massive objects like black holes, Einstein's relativity field equations prove to be much more accurate than the basic Newtonian equations.

One example of how Einstein's view of gravity works better than Newton's is the fact that gravity bends the path of light just as it does the path of matter. Light doesn't have mass, so Newton's equations would suggest light is unaffected by gravity. But gravity does bend the path of light in a phenomenon known as gravitational lensing, just as Einstein's equations predict. We will be talking more about gravitational lensing in Chapter 8.

Compared with Newton's gravity, Einstein's version is more precisely how space and time behave, and it is a better representation of what gravity really is. But we still use Newton's version of gravity for almost all everyday calculations of gravity and its effects on matter.

We sense the warping of space quite easily, and we know it as gravity. But the effect of massive objects on time is subtle and not at all obvious to us. For example, time runs faster at the top of a mountain than it does in the valley below, because gravity is a little weaker at the top of the mountain. But the difference is far too small for us to notice! So when Einstein published his general relativity field equations in 1915, many physicists were skeptical. But there were ways to go out and test this theory. It predicted that light was deflected by gravity. It predicted the existence of "gravitational waves" – ripples in space-time caused by changes in gravity that propagate outward at the speed of light.

Gradually, experimental evidence has steadily supported Einstein's conclusions:

- There were anomalies in the orbit of the planet Mercury that had been a mystery for years – general relativity explained the anomaly.

- Gravitational deflection of starlight – it was first observed in 1919.

- Gravitation red shifting of light – was first observed in 1959.

- Gravitational waves – indirect evidence came in 1974 when it explained the orbits of certain binary-pulsars. And we have now directly observed and measured them with the two LIGO (Laser Interferometer Gravitational-Wave Observatory) detectors here in the United States in late 2015! We will talk more about this in Chapter 17.

- Black holes were predicted by general relativity – their existence was confirmed in early 2000s. And now we know black holes are very common, with one at the center of most or all galaxies. We will talk more about black holes in Chapter 15.

So it took several years to validate the predictions of general relativity. But by the early 1920s, most physicists had come to accept relativity, and Einstein had become the world's most famous physicist. Today, the evidence for general relativity is completely overwhelming.

General relativity said that gravity is the result of matter warping both time and space. We have spent some time talking about both special and general relativity because both treat space as something that can be warped, stretched or compressed. It is a something - a dynamic framework, a fabric. Einstein gave us a lot to think about. But even so, scientists didn't really feel that uncomfortable with the nature of empty space until very recently.

Chapter 4 – Quantum Theory

As we continue our quest to understand the nature of empty space, we now turn our attention to another area of physics which was closely related to Einstein's study of relativity.

Just as the speed of light was creating a lot of confusion in 1900, so was light's apparent simultaneous wave-like and particle-like behavior. We've already said that Newton suspected the particle nature of light. He called it corpuscular, meaning it came in little lumps. And we also know that Maxwell had shown that light was made up of waves – electromagnetic waves.

Experiments had by this time demonstrated both the wave-like and particle-like behavior of light, but most scientists at the time thought it had to be one or the other – how could it possibly be both?

The wave-like nature of light was easily demonstrated. When you drop two pebbles in still water, you get two waves. These two waves interfere with each other, adding together in some places and

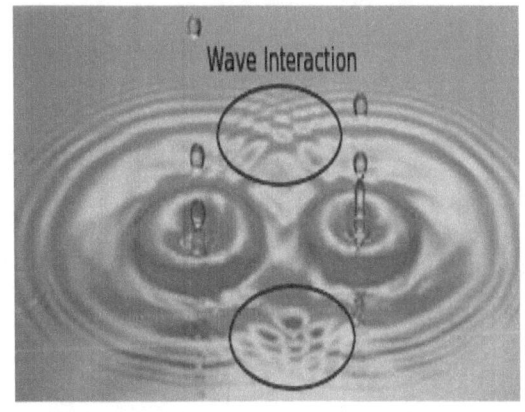

subtracting and cancelling each other out in other places. The result is called an interference pattern.

Light behaves exactly the same way. When shined though two slits onto a sheet of paper or a piece of film behind them, the same classic interference pattern appears. Clearly light behaves like a wave.

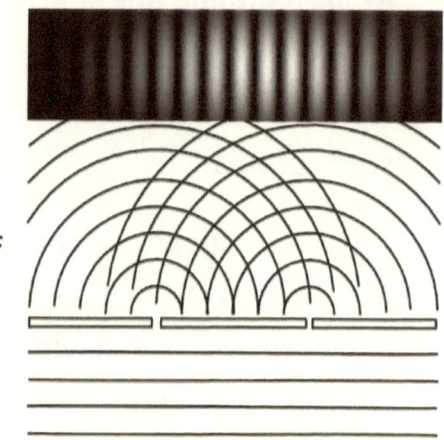

But then Max Planck came along. He was a German physicist who lived from 1858 until 1947. He suggested in 1900 that in the world of the very small, both matter and energy came only in discrete little packets which he called quanta (the plural of quantum). He was able to show in his study of "Black Body Radiation" that light was clearly being radiated in small packages of energy – what we now know as photons. Because of the way light radiated from a hot object, and specifically how its intensity varied with frequency and temperature, Planck was able to calculate the exact relationship between the energy of a single photon and its frequency. He said: $E = hf$, where E is energy, f is frequency, and h is a constant, specifically Planck's constant: $h = 6.63 \times 10^{-27}$ erg-seconds. A single photon is then an electromagnetic wave carrying a quantum of energy. Since it is a wave, it has a frequency. The higher the frequency, the more energetic that photon is.

Planck was awarded the Nobel Prize for Physics in 1918 for this discovery. He is considered the father of quantum theory. But in 1900, it still left a strange mystery regarding the nature of light. It was somehow both a wave and a particle at the same time.

In 1905, Einstein published his results on the photo-electric effect, as we have talked about previously. It provided additional confirmation of Planck's results. Light above a certain frequency could generate an electric current, while lower frequency light could not. Photons with enough energy (higher frequency) could dislodge electrons and get an electric current to flow. Lower energy photons (lower frequency) could not dislodge electrons. It was experimental evidence that supported Planck's conclusions.

In 1913 the Danish physicist Niels Bohr used Planck's constant to explain the energy states of electrons in atoms. These energy states were also quantized, with only specific energy states allowed for any given atom. He was able to explain that the frequencies of light emitted by hydrogen gas corresponded exactly with the allowed energy states of the electrons in a hydrogen atom. We were starting to realize that in the world of the very, very small, everything was "quantized", not just light!

Quantum theory continued to mature and be refined during the 1920s. There were many, many contributors.

In 1926, Erwin Schrödinger actually solved the paradox of simutaneous wave-particle behavior. He said that quantum particles propagate through space by means of probability waves. This kind of wave does not propagate through some kind of medium like water or air. It is simply an aspect of quantum particles - the quantum probability wave is the true nature of a quantum particle. It really is possible to be both a wave and a particle and neither one all at the same time!

Schrödinger went on to develop the actual wave probability equations that tell us the probability at a point in time of actually finding a particle in a particular point in space.

This is what a wave probability curve might look like and Schrödinger's equation for deriving it.

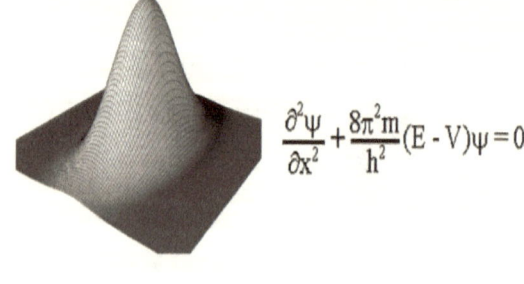

$$\frac{\partial^2 \psi}{\partial x^2} + \frac{8\pi^2 m}{h^2}(E - V)\psi = 0$$

Schrödinger and a fellow physicist, Paul Dirac, won the 1933 Nobel Prize in Physics for this work.

Werner Heisenberg in 1927 stated the "uncertainty principle", indicating the impossibility of precisely and simultaneously measuring position and energy of a quantized particle. The more accurately we know one, the less we can know about the other. And it is not because our instuments aren't accurate enough. It is because the very nature of these particles forbid that we know both at the same time. This was another way to express the "not a wave, not a particle" nature of small quantum particles.

Some scientists were deeply disturbed by the new direction quantum theory was taking physics. They didn't like probability and uncertainty replacing absolute causality. Those feelings were expressed succinctly when Einstein said "God does not play dice with the universe ".

Quantum theory continued to bring many new discoveries as well as mysteries in the second half of the 20th century. We will discuss some of them in the next chapter. But first, let's summarize some of the basic concepts of quantum theory:

1) Energy is always quantized i.e. it only comes in the form of small packets or quanta. This is a fundamental concept of quantum theory.

2) The properties of light cannot be explained by considering it to be a wave. Similarly, they cannot be explained by considering it to be a particle. They can be explained only when light is considered to be both.

3) Very small moving objects don't have a well-defined position. They only have a probability of existence at a particular time and point in space.

4) The more precisely we know the energy of an object, the less we know about its position. Similarly, the more precisely we know its position, the less we know about its energy.

5) The act of measuring the position of an object is itself the cause of it having a position. Up until then, it was only a probability wave.

As strange as quantum theory sounds, there is little doubt about its validity today. Experimental evidence that validates quantum theory has been piling up for one hundred years.

Chapter 5 – Quantum Field Theory

So let's now continue to follow the discoveries in quantum theory during the 20th century. You may be wondering at this point what any of this has to do with empty space. We will be getting to that soon, as quantum theory leads to some big mysteries about space and time.

So far, quantum theory has dealt with photons of light, and matter made up of atoms. But in the mid-1920s, some physicists began trying to apply quantum theory to force fields, starting with electromagnetic fields.

It might be helpful to talk for a second about what a field actually is. It is a physical thing made of energy that occupies space. It has magnitude and direction at every point in space. The gravitational field of the Earth points toward the center of the Earth, and its magnitude declines as you move farther away.

The Earth also has a magnetic field, which also has a direction and magnitude at each point in space. At the right is an image of the Earth's magnetic field. Similarly,

Earth's magnetic field

the strong force and

35

electromagnetic force are the fields which hold atoms together - the strong force holding the nucleus together and the electromagnetic force holding the electrons in orbit about the nucleus.

Physicists started pursuing quantum field theory while trying to understand and model how atoms radiate light as their electrons drop to lower energy states. Paul Dirac (1902-1984), who was an almost unknown physicist in 1926, published a relativistic quantum theory for electromagnetic fields. It was mathematically very complicated, and was plagued with issues like predictions of infinite quantities, but it earned Dirac the Nobel Prize in 1933 (the one he shared with Schrödinger). Dirac's theory also predicted the existence of a new particle, identical to the electron, but with a positive charge instead of negative. This particle, the positron, was later found experimentally.

A few years later in the late 1940s, Dirac's work was refined and modified by Richard Feynman (1918-1988) and others to form a new quantum field theory called quantum electrodynamics. It side-stepped Dirac's problems with infinities in a process called renormalization. It was able to produce incredibly accurate predictions of the interactions between electrons and electromagnetic fields, which were all experimentally confirmed. The Nobel Prize in Physics for 1965 was awarded to Feynman, Sin-Itiro Tomonaga, and Julian Schwinger for their work on quantum electrodynamics.

Quantum electrodynamics also produced a very strange new idea. The repulsive force that pushes two negatively charged electrons away from each other, or the attractive force that binds electrons to the nucleus of an atom, are actually a result of the particles exchanging photons! The photon is the carrier of the electromagnetic force.

The math of quantum electrodynamics, as well as the experimental results that confirm its validity, suggest some other things. Every electron is surrounded by a cloud of "virtual" photons. And empty space itself is not really empty, but rather a sea of little tiny fluctuating electromagnetic fields. These "vacuum fluctuations", as they are called, are essential in explaining spontaneous emission of light by atoms. In addition, they have actually been confirmed to exist experimentally by very precise measurements.

So once again, physicists have concluded that space is not empty!

Before we go on, we need to dwell for a moment on "virtual photons" and virtual particles in general, which we are going to talk about more in the next chapter. While historically we have been calling these things particles for the past 70 years, it tends to be misleading and in turn leads to some confusion. Fields, whether electromagnetic, gravitational, or otherwise, get disturbed by matter. In the case of an electromagnetic field, a charged particle like an electron disturbs the field. Those disturbances get sent out as ripples. But, as with everything else at the subatomic level, the ripples must be quantized. And, quantized electro-magnetic fields are what we call photons. So try thinking of these virtual particles in general as transient ripples. It is probably more accurate and descriptive than calling them particles. Now, having said that, we will follow convention and continue calling them virtual particles.

Now to conclude our brief discussion of quantum field theory, we should talk a little more about the photon being the carrier of the electromagnetic force. Today we believe that all four of the basic forces in the universe are the result of a carrier particle. These forces, in addition to the electromagnetic force, are the strong nuclear force, the weak nuclear force and gravity. The particles that carry the last three forces are known as bosons. The bosons that carry the strong and weak nuclear forces have been detected and verified experimentally. The one for gravity, the so-called "graviton particle", has not. It will be the discovery of the century if and when it is detected! We will talk more about the graviton in Chapter 16.

Chapter 6 – Quantum Foam

In the last chapter we said that empty space itself is not really empty, but rather a sea of little tiny fluctuating electromagnetic fields.
John Wheeler (1911-2008), an American physicist, took this a step further in 1955. He said that empty space was full of fields in general, and that because of the Heisenberg uncertainty

principle, which we talked about in Chapter 4, pairs of matter and anti-matter particles must be popping in and out of existence all the time. Now this would be happening in extremely short distances and for extremely short periods of time, but it would make empty space on a very small scale into a seething caldron of constantly changing fields and virtual particles.

On the surface this seems like a preposterous theory, but it was consistent with and comes about as a result of quantum field theory and quantum electrodynamics that we discussed in the last chapter. But even more importantly, in the past fifty years, experimental evidence and quantitative data have pretty much confirmed that this quantum foam is real and measurable.

The first experimental evidence for quantum foam comes from discrepancies in the magnetic field associated with the electrons in the hydrogen atom. The original experimental results were 0.1% different than what was predicted. But when quantum electrodynamics and the quantum foam theory were used to refine the prediction, it matched the experimental results out to 12 places past the decimal point – i.e. incredibly accurate.

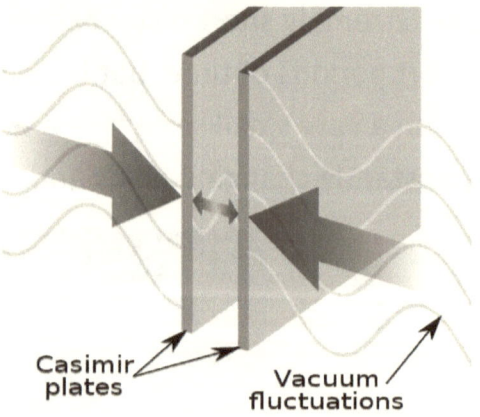

Casimir plates

Vacuum fluctuations

The second experimental evidence for quantum foam was first predicted by Dutch physicist Hendrik Casimir in 1948. He suggested that two metal plates, if placed sufficiently close together, would only allow shorter wavelength virtual particles to exist in between them. This would in turn cause there to be fewer virtual particles in between the plates than on the outward sides of them. This would then create a measurable pressure pushing the two plates together. It took about 50 years before equipment was sophisticated enough to even observe the effect, but then in 1997, it was actually measured to a high accuracy and seems to match Casimir's prediction exactly.

At this point, you might reasonably ask how light is able to travel billions of "light-years" across the Universe without any attenuation or dispersal if it has to pass through all this quantum foam. In fact, that argument does suggest that quantum foam must be happening on an incredibly small scale.

A light-year, by the way, is the distance light travels in one year. Light travels at 186,000 miles per second. So with 31.5 million seconds in a year, light travels 186,000 x 31,500,000 miles in a year. So a light-year is 5.88 trillion miles – an incredibly large distance to us, but a relatively short distance in interstellar space!

Now, when we say the quantum foam is incredibly small, what does that mean? The smallest scale possible is based on a unit called "Planck Length", which we will discuss more in Chapter 14. Planck length is essentially a mathematical construct of the minimum length where length still has any meaning. It is calculated as 1.6×10^{-35} meters or about 10^{-20} times the size of a proton. It is believed that the quantum foam and the strings in string theory (which we will talk about later) are both occurring down at the scale of a few Planck lengths.

Chapter 7 – The Big Bang

We are going to change subjects now and look at the big bang – the origin of our universe.   It will tell us some other things about the nature of empty space.

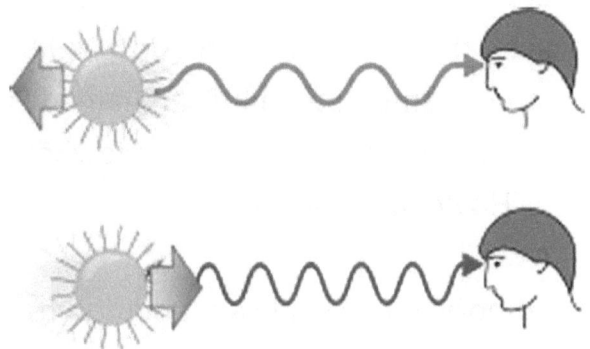

Red shift in light from a star moving away from us.

Today we know that the universe is roughly 13.7 billion years old.   One of the first and most important things that helped us estimate the age of the universe was a discovery in 1929 by Edwin Hubble (who the Hubble Space Telescope is named after) that the light coming from very distant galaxies was shifted in frequency. This phenomenon, known as "red shift", was occurring because these distant galaxies were travelling away from us at very high speeds. He also discovered that the farther away they were, the faster they were moving away. The universe is expanding, and it is doing so at a fairly high rate. And if it has been expanding rapidly for a long time, then it must have been a lot smaller in the past.

This suggested that perhaps at some point a long time ago, the universe was extremely small.   Maybe it started out as just a tiny ball of infinitely hot gas. This theory was given the name "Big Bang" in 1949 by an English astronomer and mathematician named Fred Hoyle.   The term was intended to be somewhat derogatory, as Hoyle doubted its validity.

The big bang theory got a significant addition to its credibility in 1969. With a traditional optical telescope, the space between stars and galaxies is completely dark. But mathematical models of the big bang suggested that there ought to be a residual "afterglow" - perhaps very faint after almost 14 billion years, but still out there.   And in 1969, sensitive radio telescopes first observed this faint background glow.

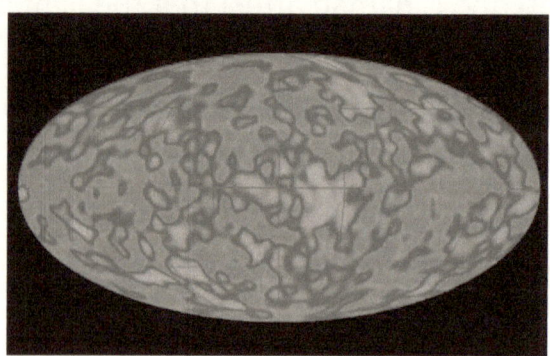

A map of the Cosmic Microwave Background Radiation created by the NASA's COBE satellite

It's known as the "cosmic microwave background radiation". It is almost exactly the same in all directions and not associated with any star, galaxy, or other object. It's another something that seems to permeate all of empty space. Its frequency and strength match exactly with the big bang's theoretical "afterglow", so it is considered one of the strongest pieces of evidence to support the big bang theory.

We believe the big bang and our universe itself started from a "singularity", a point of infinite density – just like the singularity at the center of a black hole. There was not a giant explosion – just a tiny balloon expanding to the size of our current universe.

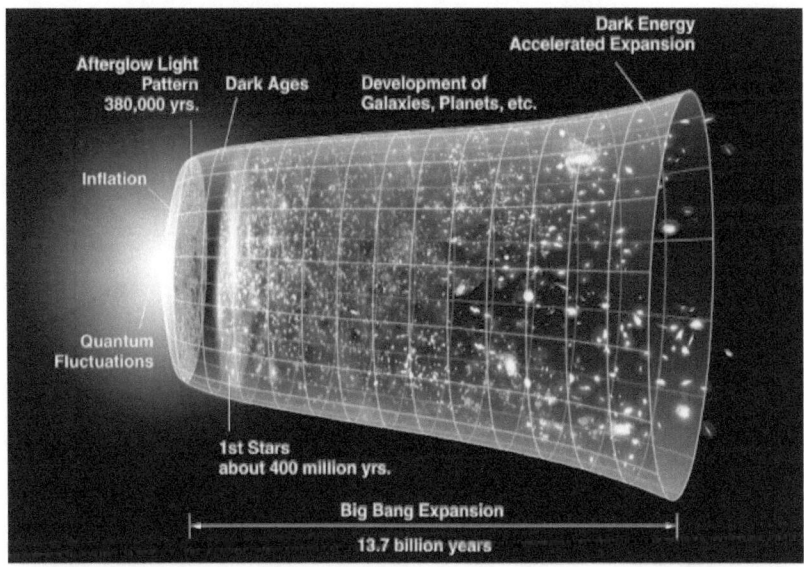

In the beginning, the universe was very, very small and very, very hot. It was so hot and so dense that matter as we know it did not exist. Light did not exist because the universe was too dense for photons to travel through it.

There was an initial period of very rapid expansion known as the inflation period. During this period, it's believed that space itself was expanding at a rate many times the speed of light.

One of the mysteries in physics today is how this period of rapid inflation was able to take place at all, i.e. why didn't gravity inhibit it, slow it down or even stop the expansion altogether? One theory says that the universe was too hot and too dense for even gravity to exist.

But this idea of space expanding faster than the speed of light is itself a very strange concept. First what is it that is expanding faster than the speed of light? And how is it able to expand faster than the speed of light? Didn't Einstein tell us that nothing moves faster than the speed of light?

Actually, space is expanding faster than the speed of light even today. The farther out into space we look, the faster distant objects are moving away from us. Simply because of the current rate of expansion, objects outside the so-called "Hubble Sphere", approximately 14 billion light-years in radius, are moving away from us faster than the speed of light, due primarily to the expansion of space itself!

Our observable universe is limited to what is inside the Hubble sphere. But actually, that is not exactly true. When we look at objects billions of light-years away, we are also looking billions of years into the past, just because the light takes so long to reach us. We can actually see objects today that were inside our Hubble sphere a long time ago, but which today, if they still exist at all, have moved way outside the Hubble sphere.

So space can expand faster than the speed of light! And it's filled with the cosmic microwave background radiation. Once again, our current understanding of the big bang leaves us guessing about the nature of empty space.

Chapter 8 – Dark Matter

Another strange set of observations occurred during the 20th century. The first was by Fritz Zwicky, an astronomer studying galaxy clusters. In the 1930s, by observing the motion of the galaxies within a cluster, he was able to deduce the amount of mass within them. By measuring the brightness of the individual galaxies, he was able to calculate the amount of visible mass within the cluster. His calculations showed that the total mass was at least ten times the visible mass! So, 90% of the mass in the galaxy cluster was some kind of mass that was hidden from view.

No one paid too much attention to Zwicky's strange observations at the time. It was thought there must be massive amounts of interstellar gas, if Zwicky's results were even valid at all. But during the 1970s, two other astronomers, Vera Rubin and Kent Ford, looking at the motion of stars in nearby galaxies, came to the same conclusion. The stars were rotating much faster than they should for the observable mass of the galaxies. And stars toward the outside edges of galaxies were rotating around its center almost as fast as the ones toward the center. This suggested that the extra mass was somehow spread out in a giant cloud around the galaxy, not concentrated in the center. The term "dark matter" was coined to describe this extra mass.

More recent observations using gravitational lensing have allowed physicists to actually map this invisible matter. Gravitational lensing occurs when light from a distant galaxy in bent and smeared out as it passes by a much closer massive object.

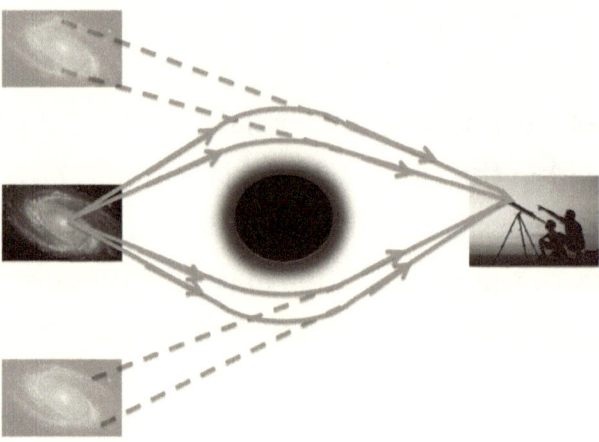

The resulting images, like the one below, show the distant galaxies smeared out in circular arcs around the mass which is distorting it.

Using computer models generated by this gravitational lensing, we have been able to generate images of dark matter and superimpose them onto actual images of galaxy clusters. In the one below, the dark matter is shown as the wispy, fog-like stuff.

But what is dark matter? It neither emits nor absorbs light of any frequency, so it's invisible, as the name implies.

It is not just vast clouds of hydrogen gas that never coalesced to form stars. Clouds of hydrogen gas do, in fact, absorb and re-emit light. And hydrogen gas would not possess enough mass to account for dark matter's gravitational effects, even if it didn't absorb or emit light.

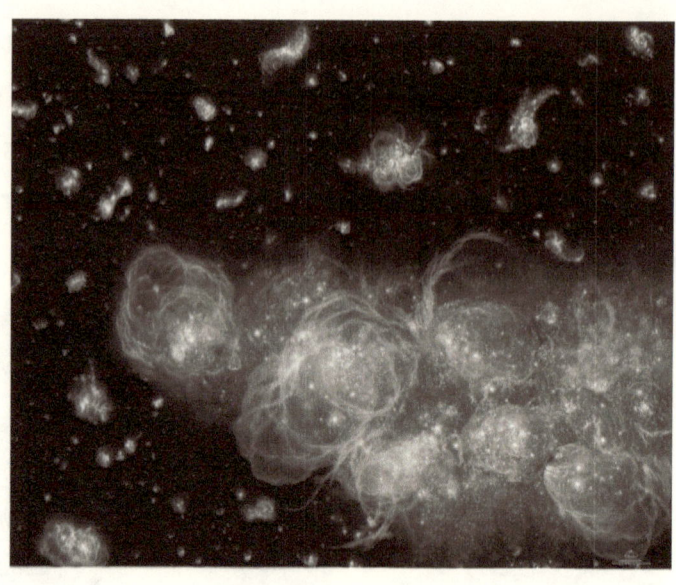

Another Image of Dark Matter

Another possibility is that dark matter is brown dwarfs, small stars that never achieved enough mass to ignite the hydrogen fusion process that powers regular sized stars. But NASA's Wilkinson Microwave Anisotropy Probe (WMAP), a satellite that operated from 2001 until 2009, determined that ordinary atoms make up only 4.6% of the universe, suggesting that dark matter cannot be made up of atoms.

So the most likely explanation for dark matter is something called WIMPs (Weakly Interacting Massive Particles). These are particles with high mass and yet have very little interaction with regular matter. But because they do have mass, they do interact with regular matter through gravity. And theoretically, WIMP particles could have been produced during the

Big Bang in sufficient quantities to account for dark matter.

Many researchers today are trying to find experimental evidence for the existence of these WIMPS. But to date, WIMPs have not been detected. So for now, dark matter remains a mystery.

Empty space contains enormous amounts of dark matter. However, it is clearly not distributed evenly throughout space. And therefore, it is not a characteristic of empty space like some of the other phenomenon we have discussed.

Chapter 9 – Gravity Probe B

In 2004, a spacecraft was launched by NASA containing an experiment that was 50 years in the making. It was called Gravity Probe B. It was placed in orbit 400 miles above Earth. Its intention was to confirm with direct experimental evidence that space is dynamic - something that can be stretched and warped, just as Einstein had predicted in 1915.

Two physicists, George Pugh and Leonard Schiff, proposed the idea in 1959 of putting a gyroscope in orbit. By carefully measuring the precession or wobble of the gyroscope's spin axis, it would be possible to measure two of the effects of general relativity.

The first is called the geodetic effect – that is the amount by which the Earth's mass warps space-time in its immediate vicinity. The second thing is something called the frame-dragging effect. It was first postulated in 1918 as a corollary to general relativity by Austrian physicists Josef Lense and Hans Thirring. They proposed that massive bodies, like the Earth, should drag local space-time around with them as they rotate.

Work continued on the design and refinement of the gyroscope in orbit experiment for about the next 40 years. It turned out to be much more technologically challenging than what had originally been envisioned. The geodetic effect was predicted to result in a gyroscope precession of 0.0018 degrees/year. The frame dragging precession was much smaller yet – 0.000011 degrees/year! These were incredibly small variations – impossible to measure with the tools available back in the 1960s.

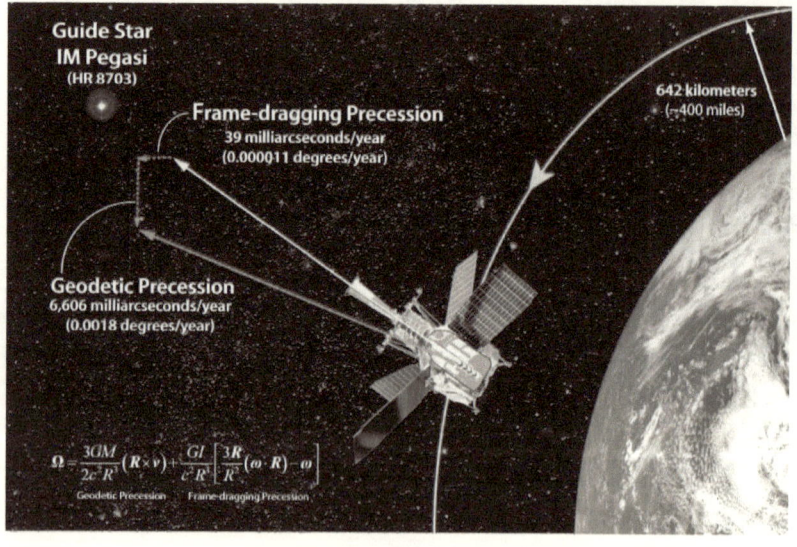

A lot of new technology would need to be developed and implemented to make the experiment accurate enough. The final version of the experiment consisted of a set of four very high precision gyroscopes all tied to a telescope that could be locked onto a distant star. The movement of the star in the telescope view frame would indicate the wobbles of the gyroscopes. After considerable refinement, this arrangement finally seemed accurate enough to measure the minute precessions or wobbles in the axis of the gyroscopes.

Actually fabricating the experiment and getting it scheduled on a NASA launch took another several years, but the experiment was finally put in orbit in 2004.

Results were to start coming back in a couple of years, but technical problems required repairs. After the repairs, results continued to come in, and by 2011, the team reported excellent correlation between the precessions predicted by general relativity and the actual data.

Gravity Probe B clearly provided experimental evidence of space being both warped and dragged by the Earth. And Earth is tiny compared to any star or black hole. You would not need sensitive instruments to observe these effects on space around a black hole. They would be violently evident!

Chapter 10 – Theory of Everything

Albert Einstein spent the last 40 years of his life working on what he called the unified field theory. He was attempting to unify the forces of gravity and electromagnetism. He never succeeded.

Physicists ever since Einstein have been trying to derive one overarching theory that explains all the forces in nature. It has gotten a lot more complicated since Einstein's time. For one thing, two other fundamental forces were discovered – the strong and weak nuclear forces. And there is one other thing that any good theory of everything, as the quest is most often called, must do. It must resolve the apparent conflicts between general relativity, which works incredibly well on very large stuff, and quantum theory, which works incredibly well on very small stuff.

Why are there conflicts? The mathematics of the two is incompatible. When Einstein's field equations are applied to small particles where the quantum nature of matter must be taken into account, the math blows up. It starts generating lots of infinities which tells physicists that something is not working correctly.

But there are places where we need both general relativity and quantum theory to work with each other. One is trying to model and understand black holes.

Black holes have a huge mass which requires general relativity to model properly, but they also have a singularity where that huge mass is crushed into a minutely small volume where quantum effects are important. Understanding black holes requires finding a way to make general relativity and quantum theory work together. We will talk more about this in Chapter 15, which deals specifically with black holes.

So our theory of everything must not only unite all the forces in nature, but must also unite quantum theory and general relativity.

One branch of physics that has shown a lot of promise for achieving a theory of everything is supersymmetric string theory, or just "string theory" for short.

String theory makes the assumption that all the fundamental particles are actually tiny vibrating loops (or strings) of energy built in 10 dimensional space-time. Since we live in 4-dimensional space-time, the assumption is that the other 6 dimensions needed to make string theory work must fold in on themselves in a small enough space that they aren't visible to us.

String theory has one huge thing going for it. It is able to theoretically predict all the particles in the Standard Model, and all of their characteristics. This is quite an amazing feat.

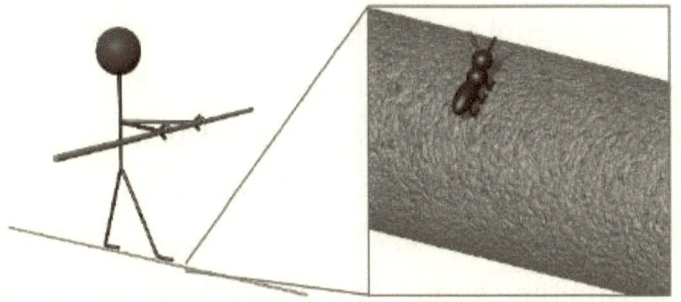

Above is an illustration of how the extra dimensions in string theory might exist without our being able to observe them. In our world, the tight-rope walker can only move in a single direction. But on a much smaller scale, an ant sees the tight-rope as having two dimensions, one of which is a small closed loop.

The "Standard Model" is where we have finally arrived at all of the smallest particles – the quarks that protons and neutrons are made of, the leptons like the electron, and the bosons that enable the four fundamental forces in the Universe.

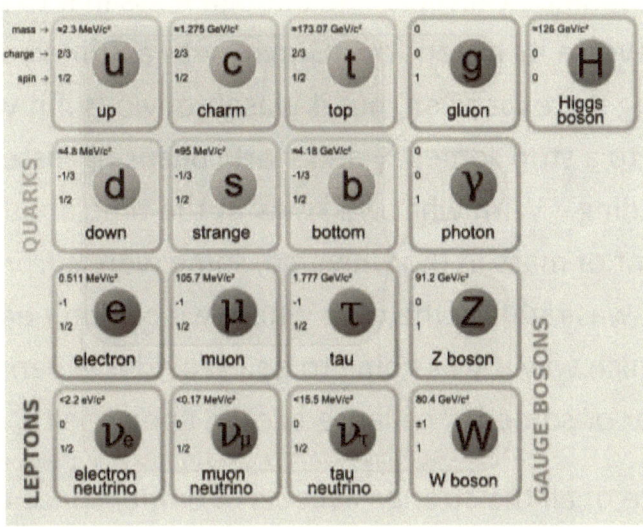

The Standard Model

Again, string theory predicts all the particles in the Standard Model and all their characteristics. But string theory also has one huge negative. String theory is just a very beautiful mathematical construct. Because the strings are so small, we don't have any tools with which to experimentally verify their existence. Nor is it likely we will anytime soon. Indirect evidence to support string theory will come as new discoveries continue to fit with string theory, but it will probably remain an unproven theory for many years to come.

Nevertheless, by 1990 or so, string theory had made a lot of progress, and physicists felt pretty good about their collective progress. It seemed we were closing in on the theory of everything.

In cosmology, the big mystery at the time was whether the expanding Universe had enough mass in it to slow and reverse its expansion. Gravity was assumed to be slowing the expansion, but it wasn't obvious if it would come to a stop someday and start collapsing instead of expanding. With what we knew at the time about the amount of mass in the Universe, some were surprised that it was right on the cusp where we couldn't easily tell which way it was going to go. Would it expand forever or someday collapse back in upon itself?

In spite of all the strange aspects of empty space that had been observed over the past 150 years, including dark matter, as the 20th century drew to a close, physicists and cosmologists felt they were getting close to understanding the universe pretty well. We still had to resolve the conflicts between relativity and quantum theory, but progress was being made. Physicists were predicting that a theory of everything was at hand and that, perhaps in the next 25 years of so, there would be nothing left to learn.

But all that changed, starting in 1998, when we made a major discovery in cosmology. It instantly meant we still have a lot to learn, especially about empty space. The theory of everything was going to have to wait a while.

Chapter 11 – The Accelerating Expansion of the Universe

As we said in the previous chapter, the big mystery in the early 1990s was how fast the expansion of the universe was slowing down. As shown in the image below, the question was if the expansion was slowing down fast enough to eventually start contracting into a "Big Crunch", or would it continue to expand forever in some fashion?

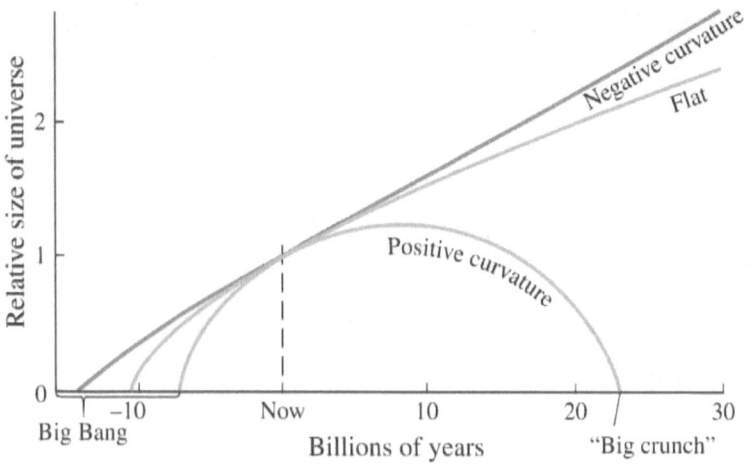

In 1998, two teams of astronomers were studying Type 1A supernovae in an effort to find out how fast the expansion of the universe was slowing down. The two teams were the High-Z Supernova Search Team and the Supernova Cosmology Project. The assumption was that the expansion of the universe must be slowing down because gravity should be gradually trying to pull everything back together. All these teams were trying to do was find out how fast the expansion was slowing down.

A Supernova is a gigantic explosion of the massive star – it can outshine an entire galaxy for a short period of time.

Type 1A supernovae are a type of exploding star that does so in a very predictable way, producing a very specific and predictable brightness. For a few days that brightness is equal to all the stars in a galaxy, so we can see them from billions of light-years away. And because we know exactly how bright they really are, using their perceived brightness, we can measure how far away they were.

We also have talked about how Edwin Hubble was able to measure "red shift" in the frequency of light coming to us from a star. From the red shift, we can calculate how much the light waves are stretched and therefore how fast the star was moving away from us when the light left the star.

Using both these pieces of data from a number of Type 1A supernovae, the teams attempted to calculate how fast the expansion was slowing down over time. They came up with a totally surprising, almost unbelievable result. The expansion wasn't slowing down at all. In fact, it was accelerating!

The image below shows the data. The straight line is what you would expect if the universe was expanding at a constant rate. Instead the data itself is lining up above the straight line, indicating that the expansion is accelerating.

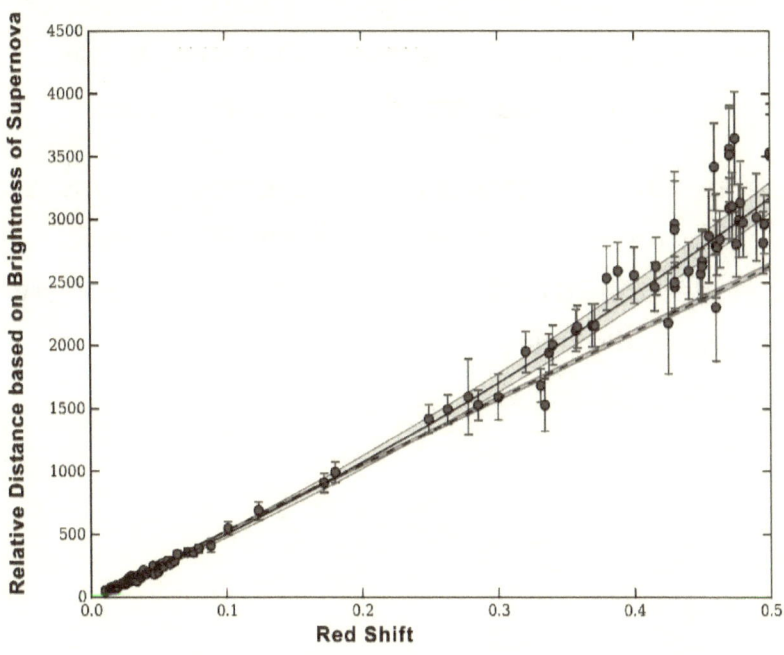

So the graph showing the possible options for the expansion of the universe needed to be amended. A new line had to be added showing that the expansion was accelerating!

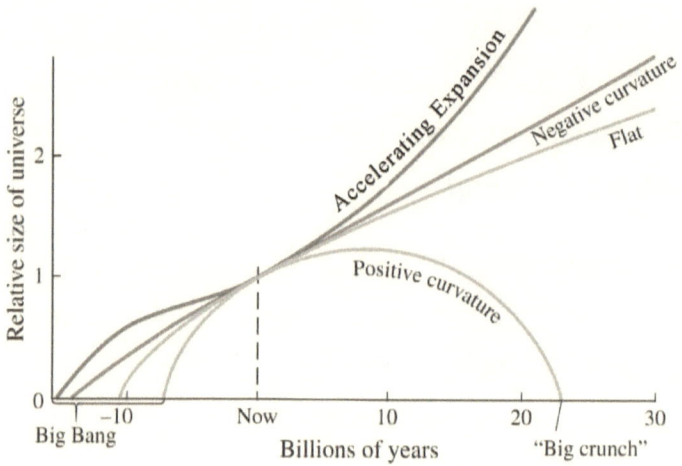

This new discovery turned physics and cosmology on their heads. It meant there was something totally wrong with our understanding of the universe.

Three individuals from the two supernovae teams (Saul Perlmutter, Brian Schmidt, and Adam Riess) were awarded the 2011 Nobel Prize in Physics for this very surprising discovery.

Chapter 12 – Dark Energy

So now we have to figure out why the expansion of the universe is accelerating. Some mysterious new form of energy is creating pressure on space itself to cause it to expand at an ever increasing rate. This energy has been given a name – "dark energy". But it is just a name. We don't know what dark energy actually is, but it seems to be a characteristic of space itself.

While we don't understand what dark energy actually is, we do know how much dark energy is required to produce the force we see – the one that overcomes gravity and is causing the universe's expansion to accelerate instead of decelerate. Assuming the equivalence of mass and energy using Einstein's $E = mc^2$, this dark energy equals roughly 70% of all the energy in the Universe!

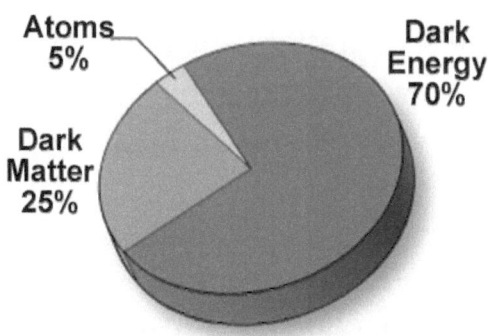

Makeup of the Universe

If that isn't strange enough, the ordinary matter and energy that makes up all the stars and planets, all the galaxies and black holes, all the matter and energy (like light) that we can interact with, only makes up another 5% of the energy in the universe. The remaining 25% of the universe is made of the dark matter which we discussed in Chapter 8.

Now you may be wondering how dark energy can possibly be 70% of the total energy in the universe and yet be so unapparent that we only recently became aware of its existence.

Dark energy is a characteristic of empty space. The density of dark energy is very, very low. Its huge total comes about because of the enormous amount of empty space in the Universe. Think of the incredible distances between galaxies, typically millions of light-years. And even within a galaxy, there are typically several light-years of empty space between each star! We can calculate the energy density required to generate the accelerating expansion of the universe. And when we multiply it by the enormous amount of empty space in the universe, it produces a gigantic number.

But what exactly is dark energy. We just don't know. There are a bunch of theories. One says Einstein's general relativity must be wrong after all, but there is so much evidence that it is right, that seems unlikely.

Another theory says that those virtual particles that spring momentarily into existence could be the source of dark energy. As we said in the chapter on quantum foam, we can even measure a pressure produced by these virtual particles. But if that is the source of the accelerated expansion, its calculated value would be many orders of magnitude too high. At present, it doesn't seem to account for dark energy.

Yet another theory, and probably one of the more interesting ones, says that dark energy is the result of a fifth fundamental force in nature. It's been given the name "quintessence". This new force must have some strange properties to explain the phenomenon of dark energy. Its strength has to vary depending on how much matter is in its vicinity. The force gets weaker as the amount of matter nearby increases. So it would be very hard to detect here on Earth. In the empty regions of space, however, the force would become powerful enough to push the matter in the universe apart – the opposite effect of gravity.

Or course, if quintessence is a fifth fundamental force, like the other four, as we learned in our chapter on quantum field theory, it must be quantized and therefore have an associated particle. That particle has also been given a name – the chameleon particle.

The search is on to find the chameleon particle. Several clever experiments have been devised to look for it, and they are reasonable lab experiments, not requiring a massive multi-billion dollar particle accelerator.

At the present time, however, there is no experimental evidence for either the quintessence field or the chameleon particle. Dark energy currently remains a mysterious property of empty space!

Chapter 13 – So What Do We Know?

The last 12 chapters have given us some historical perspective.   Today we know a lot about the Universe and about empty space. In this chapter, we will summarize what we've learned.   Then, in the remaining chapters, we'll move into some areas where we still are trying to learn!

We saw in Chapter 1 that light always travels at the same speed through empty space, no matter what its source or frame of reference. In Chapter 2, Einstein explained this by telling us that space and time are not fixed standards, but both can change depending on our frame of reference. Then in Chapter 3, Einstein told us that gravity is actually a result of empty space being warped by the matter in it. Specifically, space contracts and time slows in the presence of matter! Space itself is strange stuff.

In Chapter 4, we looked at quantum theory and the dual wave-particle behaviors of light. We found that all forms of energy and matter are quantized when we get down to very small scales. Then in Chapter 5, we looked at quantum field theory and found that electro-magnetic fields are also quantized, and that, as a result, empty space in teaming with particles that make up the fields within it! And then in Chapter 6, we looked further into the quantum foam of particles that is empty space. We saw experimental evidence that supports the existence of this quantum foam.

In Chapter 7, we looked at the big bang which created the universe. We saw that the cosmic background radiation left from the big bang still permeates all of space even today. And we learned that space has been able to grow and expand even faster than the speed of light. In Chapter 8, we learned that there is more mass in the Universe than we previously thought – a lot more, hidden from our view. Then in Chapter 9, we learned about Gravity Probe B which produced direct experimental evidence that space can be warped and dragged even by our own small planet.

In Chapter 10, we learned about the theory of everything – how, in the late 20th century, string theory and our understanding of the universe were starting to make us think we really had everything figured out. But then in Chapter 11, we found out that the expansion of the Universe is actually accelerating. It was clear we still had a lot to learn. It was clearer than ever, that we don't understand the nature of empty space! And finally, in Chapter 12, we looked at the force that is accelerating the expansion – dark energy, which still remains mostly a mystery.

So where does that leave us? How do we learn about the nature of dark energy? How do we get back on track in our quest for a theory of everything? And how do we finally figure out what the real nature of empty space is? That is the subject of the rest of this book!

Chapter 14 – Quantum Space and Time

As we have talked about in previous chapters, we learned in the early 20th century that both matter and energy come in small packets called quanta. By the middle of the 20th century, we learned that electric and magnetic fields were also quantized. We came to believe that all fields, including gravity, are quantized, though we don't yet have a complete theory for quantum gravity.

Quantum gravity was first proposed about 70 years ago, but it was a strange concept and seemed to conflict with Einstein's view of gravity as a distortion of space. But the more we learn, the more physicists have come to believe that gravity must be quantized. Today, quantum gravity is one of the greatest pursuits in physics. Various quantum gravity theories are now in competition with string theory to see which will claim the grand prize of becoming the theory of everything!

We will talk a lot more about quantum gravity in Chapter 15, but for now, we will first explore one basic tenent of quantum gravity – the idea that everything is quantized, even space and time!

To get our arms wrapped about this concept, let's start with what we all know. Both time and space appear to be continuous and smooth to us. They don't come in little chunks that can't be subdivided, right? We know that time can be subdivided to well below a billionth of a second, because our existing computers can actually perform operations in fractions of a billionth of a second. And we know that space can be subdivided down to atoms, and beyond that to the subatomic distances within the nucleus of atoms, and eventually, down from that to the dimensions of the strings in string theory from which we think elemental particles originate. So time and space are smooth and continuous to many orders of magnitude below our everyday experiences.

But does that mean space and time can't be quantized. Actually, thanks to today's digital electronics, we have everyday experience with things that appear to be smooth and continuous but are actually totally quantized – quantized in the sense they are made up solely of computer digitized data – just ones and zeros!

So here is a picture of the pyramid at the Louvre in Paris. But now let's look at the top of the pyramid close up. What do we see? The picture is actually made up of pixels – tiny discrete squares of a single color, not smooth and continuous at all.

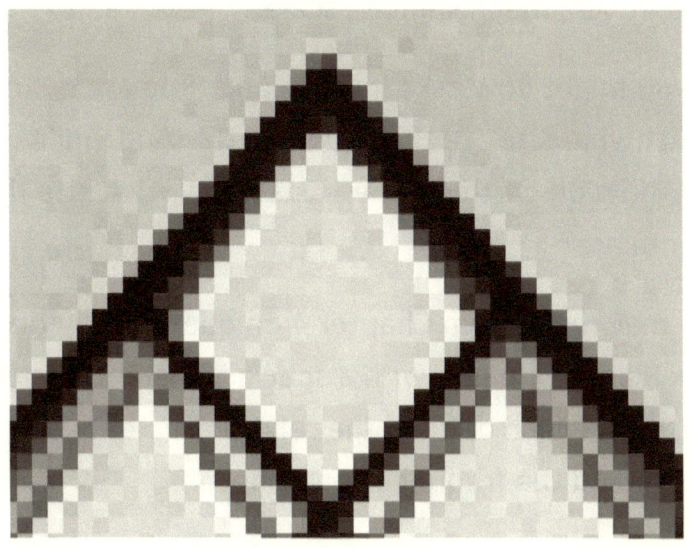

So this is an example where space has been quantized on a fairly large scale. But even so, unless we examine it with a microscope, we see nice smooth surfaces and continuous space.

What about time? Again we can turn to digital electronics for an example. A music CD samples a soundwave of our music 44,000 times per second, and records its amplitude as a number. It has literally chopped time up in little chunks, each $1/44000^{th}$ of a second. Yet it can put these chunks back together, and we hear the result as continuous, smooth sounds!

So if we can do this on a relatively large scale with our electronics, it should not be too surprising to find out that nature itself is doing much the same thing - except for being on a much, much smaller, or in the case of time, much, much shorter, scale.

So you might now ask: if space and time are quantized, at just what size are they quantized? And while we're at it: how on earth do physicists know what size these quanta are?

You may recall from Chapter 4 on quantum theory that Max Planck came up with a special constant, the Planck constant, which defined the relationship between the energy and the frequency of a photon.

He also proposed a special distance associated with this constant, an incredibly small distance, which has come to be known as the Planck length. It arises naturally over and over again in the math that seems to govern quantum mechanics.

$$\text{Planck length} = \sqrt{\frac{\hbar G}{c^3}} \quad = \quad 1.616 \times 10^{-35} \text{ meters}$$

In this equation c is the speed of light, G is the gravitational constant, and $\hbar$ is the Planck constant. To see how small the Planck length is: it's about 10^{-20} times the diameter of a proton.

The math in quantum gravity theories suggest that the Planck length has special meaning as the shortest possible distance that has any meaning. Similarly, a special unit of time, called the Planck time, is defined as the time required to travel the Planck length at the speed of light. It is believed to be the shortest amount of time that has any meaning. The Planck time is about 5×10^{-44} seconds.

These are both incredibly small numbers, but physicists now believe that space and time are likely themselves quantized at these dimensions. That might seem to imply that space is composed of three dimensional pixel-like elements that are of 1x1x1 Plank length dimensions. But at these dimensions quantum behavior totally dominates. So it more likely suggests space itself is a soup of minute space particles. It may suggest that these space particles form a textured framework in which everything else happens. In loop quantum gravity, which we will discuss in Chapter 16, this textured framework comes in the form of a network of loops.

At the very least, all of this again suggests that we have a lot to learn about the true nature of space.

Chapter 15 – Black Hole Mysteries

In this chapter we will talk a little more about black holes and some of their mysterious properties. They are of interest to us not only because they are strange beasts that push the extremes to where time stops and light can't move. But they also have some interesting characteristics that shed light on the nature of empty space!

In the early 20th century, black holes were just a mathematical anomaly that came out of Albert Einstein's general relativity field equations. His math said that an object greater than 1.4 times the size of our own sun was capable of collapsing to the point where nothing stands in the way of gravity – a point at which the entire mass would collapse down to a single point. As you approach that point, gravity gets so intense at some point that even light cannot escape.

The single point of collapsed matter in a black hole is called the "singularity". The distance from the singularity where light is no longer able to escape is a sphere surrounding the singularity called the "event horizon".

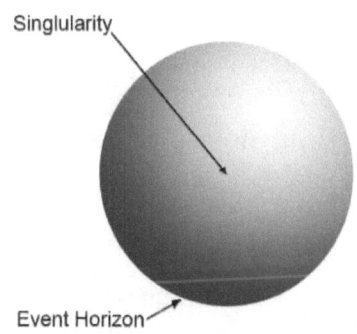

Singlularity

Event Horizon

All of this was just speculation and a subject for science fiction up until very recently. While it has seemed likely that black holes exist, proof beyond a reasonable doubt has only come in the 2000s. Today, we know that black holes are real, that they are very common, and that there are massive black holes at the center of most galaxies, including our own Milky Way galaxy. Black holes are fascinating to scientists for a number of reasons. All the rules of physics go out the window at the event horizon. Let's look at some of the strange characteristics of black holes.

Nothing, not even light, can escape. So the area inside the event horizon is, for all practical purposes, no longer a part of our Universe. All information about the matter which falls into a black hole is lost forever. However, as you will see later on in this chapter, quantum theory calls some of this into question.

According to Einstein's relativity, time is affected by gravity. At the event horizon, not only is light no longer capable of escaping, but time comes to a stop. If you were falling into a black hole, you wouldn't see time slowing down; you would find yourself falling faster and faster, just as you would expect. But to an outside observer watching from afar, you would fall slower and slower, and never quite reach the event horizon.

Gravity is an incredibly powerful force as you near the event horizon of a black hole. If you were in space and falling toward the earth, you would feel weightless. But falling feet first toward a black hole, gravity is so intense, that it might be pulling on your feet with 10,000 lbs. more force than its pulling on your head. For this reason, everything gets stretched out into long strings as materials approach a black hole. This phenomenon has the not-so-scientific whimsical name: "spaghettification".

Above is a model of a rotating black hole. In the central region, hot gas and dust is slowly spiraling into the black hole. The magnetic field close to the black hole accelerates some of the matter away, forming two high-speed jets. The grid superimposed on the image is a model of warped space-time —the fabric of space and time severely warped in the vicinity of a black hole.

The objects that form black holes, like stars, typically are rotating, like our Earth does. As they collapse into black holes, that rotation gets faster and faster, just like a skater when they pull their arms in. So it is highly likely that most black holes have a high rotational speed associated with them. The mathematics of rotating black holes get really complicated, but they seem to suggest that something going in a black hole at one point in time and space might perhaps re-emerge in a different time and /or place. This line of thinking is what has inspired all kinds of science fiction concerning "wormholes", time travel and faster-than-light movement across the universe. However, we have no way today to experiment on what actually would happen.

We only know of one way to form a black hole today. That's where a large star collapses forming a black hole that is very massive. The event horizon of such a black hole has a radius of perhaps 500 miles or more. But there is nothing in theory that says black holes have to be that big. Some physicists have argued that a large number of small black holes could have been created during the big bang. They also believe that some of the small black holes created during the big bang may be still around today. These are called "primordial black holes", and they might be as small as an atom, though trillions of times more massive.

This brings us around to another strange characteristic of black holes. We think of black holes as the ultimate trap. Nothing gets out – not even light. But there is a strange twist. Due to quantum mechanical effects, black holes leak a little bit. Big black holes don't leak much, but small ones do, and eventually, if they get small enough, they start to give off large amounts of radiation. This radiation, called "Hawking radiation", was proposed the famous physicist, Stephen Hawking in 1974. This radiation means that a black hole (that not even light can escape from) somehow leaks matter due to quantum effects. If you are a Star Trek fan, you may remember that the Romulans' warp drives were powered by a quantum singularity, i.e. an artificial primordial black hole.

By now you are probably saying "yes, black holes are interesting, but what do they have to do with our understanding of empty space?" It turns out black holes seem to provide a lot of evidence in support of quantum space and gravity!

In classical non-quantum physics, Einstein's general relativity says everything in a black hole disappears into the singularity, a point with no dimensions and infinite density where physics just doesn't work anymore. The matter that disappears into the singularity is gone forever, in every possible sense.

Hawking radiation, however, seems to suggest that that, for quantum mechanical reasons, the matter is not completely gone. It radiates slowly out from a black hole. So we have a paradox: matter that is supposed to be gone forever really isn't gone forever! This paradox has puzzled physicists for several years. It forces us to pose the question: does quantum theory also give us a reason to dismiss the existence of the singularity?

Recent studies using loop quantum gravity, which we will talk about in the next chapter, suggest the quantum nature of space precludes the existence of the singularity. Instead, the matter inside a black hole is compressed into a very small but finite volume where the laws of physics still apply, and where it is not gone forever. It resolves the paradox presented by Hawking radiation and makes the infinities of a singularity go away.

So the behavior of black holes gives us some indirect support for the quantum nature of space and for quantum gravity. But that's not all. A couple of other oddities that come out of the mathematics of black holes also give us some tantalizing hints about the quantum nature of space.

One way to deal with matter falling into a black hole is to treat it as information lost. The information in this case is the mass, momentum and spin of every particle that falls into the black hole. Strangely, the surface area of the event horizon sphere turns out to be exactly large enough to store that lost information, assuming the size required to store an individual bit (a 1 or a 0) of information is a 2 Planck length square. Proof that space is quantized? No, but probably not just a coincidence, and certainly intriguing! A young physicist in the 1970s, Jacob Bekenstein, generalized this concept and formalized it with a mathematical proof. Today it remains one of the strongest arguments for space being quantized.

One additional strange theory has come out of the Bekenstein proof. If the surface area of a black hole can contain all the information about what's in it, his math also suggests and any three dimensional space can be projected from a two dimensional surface.   This has led some to suggest a kind of holographic reality for the entire universe. Everything in it, from us to all the galaxies could simply be mere projections from a sphere surrounding our universe. Some theorists have even suggested this is very likely a better view of reality.

And here is another oddity that comes out of black hole math. A Canadian physicist, Bill Unruh, working to combine relativity with quantum theory, discovered in the 1970s that anything which is accelerating must find itself embedded in a hot gas of photons, with the gases' temperature proportional to its acceleration. This is to say that the space itself starts to emit photons in the presence of an accelerating body.

Now, as you would expect, under the normal amounts of acceleration that we encounter here on Earth, the amount of radiation is far too small to be measurable. But remember that Einstein said the pull of gravity is, in every way, indistinguishable from acceleration. If you were very near a black hole, and using powerful engines to prevent yourself from falling into it, this would require tremendous acceleration and the space around you would start emitting massive amounts of radiation.

Once again, we have a result that suggests empty space is something very special!

Chapter 16 – Quantum Gravity

The theory of quantum gravity is pretty much synonymous today with the theory of everything. It is the search for a way of finally merging quantum theory with Einstein's general relativity. It needs to be a mathematical and conceptual framework in which the two can coexist and function together. And if possible, it should be experimentally verifiable.

We don't have a single theory of quantum gravity that fully satisfies these requirements. What we have instead in a bunch of quantum gravity theories that all in some ways meet these requirements and in other ways fall short.   They approach the problem very differently, asking different questions, and result in different ways of viewing reality.   But they all have some things in common. They all tell us that gravity must be quantized, that there must be a particle associated with gravity which we call the graviton. They all suggest that space and time are probably quantized as well.

So, in this chapter, we will discuss the various approaches: how they compare and conflict with one another. The final answer will likely not be that one of these theories is right and the others are wrong. They are all looking at the same reality and trying in their own way to describe it. When one finally emerges that gives us the best description of gravity, it will probably show that many aspects of each theory are basically valid.

Let's start with string theory, which we already discussed briefly in Chapter 10, when we first introduced the concept of a theory of everything.

String theory treats all particles and forces, including gravity, as minute strings of energy. The vibrations of these strings occur in ten-dimensional space-time, with each 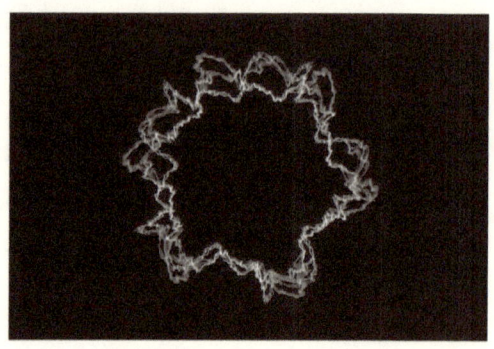 one-dimensional point in our ordinary space actually consisting of a complicated geometrical structure in six dimensions. All of these six dimensional structures exist at the Planck length scale, so they are invisible to

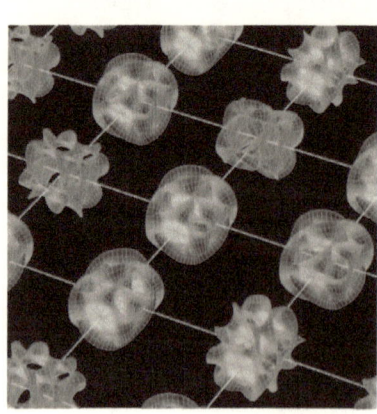

us. The vibrations of these tiny threads of energy are what replace particles and fields in the quantum description of the universe. The strength of these vibrations is what we see in our world as mass. The patterns of these vibrations are the fundamental forces.

In string theory, gravity is a field carried by the graviton particle – a specific kind of string. It is consistent with general relativity in two ways. First, both string theory and general relativity suggest that the graviton must be a massless particle with a spin of ±2. And general relativity, the Einstein field equations, can actually be derived from string theory using a process called "compactification", where the 6 extra spatial dimensions in string theory are reduced out of the math.

There are problems with string theory, however. For example, it predicts a lot of new particles which haven't yet been found. During the late 20th century, between 1985 and 2000, string theory was making great progress, and was seen as the most likely road to understanding both quantum gravity and a theory of everything. But it has made very little progress in recent years, causing many physicists to think it may not ultimately be the solution we are looking for.

And another huge problem for string theory is that it is frustratingly abstract and theoretical. It is all in the math! There is nothing about string theory that you can take into the lab and test. There are also several different versions of string theory competing with each other. Each has certain strengths and weaknesses!

Now, let's look at another theory for quantum gravity – loop quantum gravity. Here we start with an assumption that space itself is quantized. Space is woven out of loops, which is where it gets its name.

The structure of space is a network of loops called a "spin network".

One feature of loop quantum gravity is that it is background independent. That's generally considered a good thing as the theory does not rely on space being a fixed background onto which the theory is overlaid. String theory in contrast is background dependent.

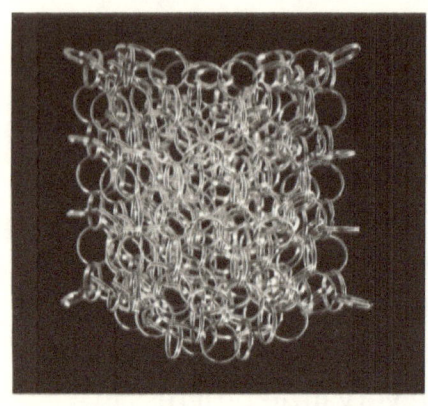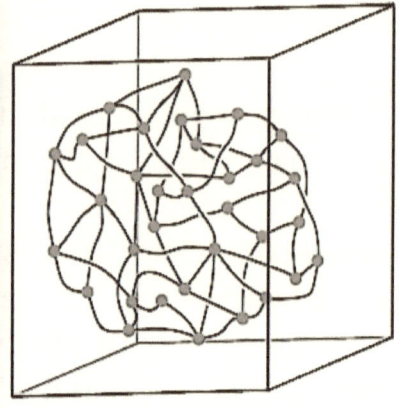

Loop quantum gravity is all mathematics, just like string theory. But it starts from very different assumptions and is therefore fundamentally different. So, in loop quantum gravity, space is a spin network with lines and nodes, as pictured above. Thus, space acquires a grainy, discrete structure – and so does time.

In simplified models of loop quantum gravity, used for cosmological explorations, it turns out that there is no big bang singularity; instead, the universe's history can be traced infinitely far back into the past, step by step, to times preceding the big bang!

Another theory of quantum gravity is called "asymptotically safe gravity". It uses some techniques derived in quantum field theory and applies them to gravity. It has had some success and some problems.

"Causal dynamical triangulations" and something called "emergent gravity" are yet two more quantum gravity theories, which have been proposed in the last 15 years.

We don't know which quantum theory of gravity is correct, or if any of them are. However, we do have a pretty good idea today that gravity is quantized, and that we will find a way to properly combine general relativity with quantum field theory. And it seems very likely that both space and time are quantized at Plank scale dimensions!

Chapter 17 – Gravitational Waves and Gravitons

We are still not finished talking about gravity. It seems that the whole notion of empty space is intertwined with the concept of gravity. So, in this chapter, we will discuss two aspects of space and gravity. At the macroscopic view of gravity, we will look at gravitational waves. Then, at the microscopic quantum level of gravity, we will talk some more about the graviton particle.

Einstein predicted gravitational waves as part of his general relativity theory. It has been 100 years since he predicted their existence. And now, they have actually been detected. An instrument known as LIGO (Laser Interferometer Gravitational-Wave Observatory), shown below, detected gravitational waves in late 2015!

Gravitational waves are the compression and expansion of space itself. They are similar to the compression and expansion of air that we know as sound waves. We have finally come full circle from our search 130 years ago for the aether that was suspected of carrying light. While there wasn't any aether, we now find that space itself can in fact transmit waves – not light waves, but gravitational waves!

A highly exaggerated illustration of gravitational waves distorting space itself.

The waves detected in 2015 were the result of two very distant, massive black holes circling each other and then finally colliding. It is not that this is the only thing that creates gravitational waves. But a cataclysmic event like that creates a massive distortion of space and the resulting gravitational waves are large enough for us to detect.

Why did it take 100 years for us to detect gravitational waves? Even a massive event like the collision of two black holes creates very small distortions when viewed from a great distance. The LIGO instruments can detect distortions of space the size of a proton! And as larger, even more sensitive instruments are built in the future, we will be able to detect gravitational waves from cosmic events much smaller than two colliding black holes.

What do gravitational waves mean to us as we try to understand the nature of empty space? Einstein already told us that mass distorts space and time, so actually detecting gravitational waves doesn't tell us anything dramatically new, but it does give us another piece of real experimental evidence that space itself can be stretched and compressed.

Now let's turn back to the quantum world and talk more about the graviton. As you recall, quantum field theory predicts that every force field is enabled at the quantum level by the exchange of a force-carrying particle. The photon carries the electromagnetic force, gluons transmit the strong nuclear force, and the W and Z bosons carry the weak nuclear force. Physicists have long predicted that gravity must also have a particle associated with it. It has never been found, but it has been given the name graviton.

Surprisingly, considering the graviton has never been observed, we already know quite a few things about it. Since gravity weakens as one over the square of the distance between two objects, the graviton must have zero mass. If the graviton did have mass, it would change the exponent of 2 in the square of the distance to something else. So the graviton is massless, like the photon, and because it is massless, the graviton travels at the speed of light!

General relativity also tells us something else about the graviton. The Einstein field equations describe matter and energy in a way that requires the graviton to have a spin of 2. Not only that, the graviton is the only possible massless particle with a spin of 2. So, if we can find a massless, spin two particle, we will have found the graviton!

With all that going for us, why is the graviton so hard to find? The answer is that gravity is an extremely weak force, and the force transferred by a single graviton is incredibly weak. Let's examine how weak gravity is by comparing it with the electromagnetic force. A one ounce magnet can easily pick up a paper clip against the force of gravity. It's a one ounce magnet pulling one way and the entire earth pulling the other way! The earth weighs 2.1×10^{26} ounces, so for the magnet to win, the electromagnetic force needs to be many orders of magnitude stronger than gravity.

And what does that say about the graviton? For the whole earth to interact with our paper clip there must be an incredible number of gravitons exchanged. Yet, if the total resulting force is so small, the force conveyed by a single graviton is very small. This implies an incredibly small interaction with matter in general, making it very, very hard for us to detect.

That does not necessarily mean that the graviton will never be found. Many physicists today are working on exotic experiments hoping to tease the graviton out of hiding. If anyone can actually find the graviton, it will undoubtedly be one of the great discoveries of the 21st century!

What does the graviton tell us about empty space? One thing is pretty obvious. Anywhere near massive objects like stars and planets, empty space must be filled with an incredibly high density of gravitons, which in turn are shaping and distorting the space they fill.

Chapter 18 – The Higgs Field

In this chapter, it may seem that we are going off on a tangent. But when we are done talking about the Higgs field and the Higgs boson, we will have one of our best ideas yet about what empty space really is!

In the early 1960s, quantum field theorists were having problems with why particles have mass. Some of the math suggested that all particles should have zero mass and travel at the speed of light, like photons. There was obviously a problem, because we already knew that all the particles that make up everyday matter do have mass.

To understand the issues here, it may be worthwhile to talk briefly about what mass is. We usually talk about mass and the force of gravity almost interchangeably, but in reality they are very different things. An object with a mass of 1 kilogram weighs 1 kilogram here of Earth. That's because we have chosen the units of mass to correspond to the weight of that mass here on Earth. But its mass doesn't really have anything to do with weight or gravity. If we took that mass to the Moon, its mass would still be 1 kilogram, but it would weigh $1/7^{th}$ kilogram. Mass has to do with inertia – how much does something resist moving, and once moving, how hard is it to stop.

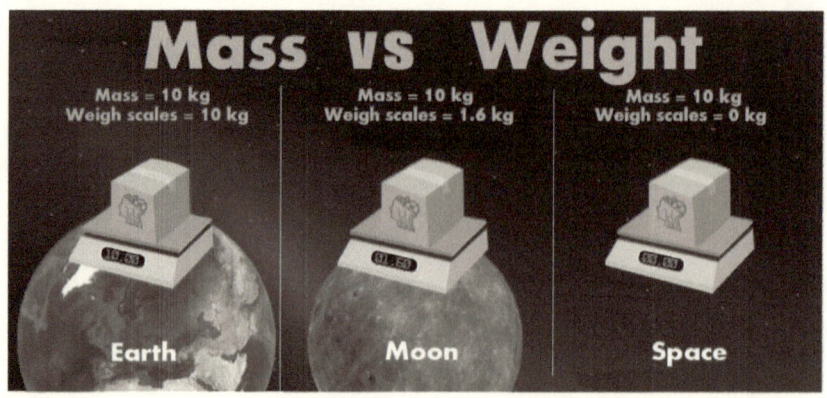

A good example of what we are talking about can be illustrated by imaging you are on the International Space Station in a truly weightless environment. You and a fellow astronaut engage in a game of catch, first with a tennis ball, and then with a bowling ball. Though both balls are weightless, you can readily imagine that playing catch with the bowling ball would be much harder than with the tennis ball, even though both are weightless. With the bowling ball, it would require a lot of work to get it moving, and then be equally difficult to stop. That's because of its mass, not its weight as a result of gravity! Without mass, everything would be zinging around at the speed of light, just like the massless photon.

In 1964, British physicist Peter Higgs proposed a new theory to explain why most common particles have mass. He proposed a new field which fills all of space. We now call it the Higgs field. This field is different than all the others. For other fields, the lowest energy state or rest state, in the absence of anything else, is 0, meaning the field is basically absent. But the Higgs field is omnipresent. It fills all of space in the absence of anything. Higgs further postulated that if the Higgs field existed, like all the other fields and forces, it should have a particle associated with it. That is again because fields are quantized, and so at very small scales, they take on the behavior of quantum

Peter Higgs

particles. So Higgs' theory included postulating the existence of the Higgs boson – a particle with the power to convey mass to other particles.

At first, Peter Higgs' theory was rejected by his peers. It took a few years, but gradually Higgs' ideas were accepted as highly likely, although unproven until very recently.

How does the Higgs field and Higgs boson give other particles their characteristic mass? Again, the Higgs field occupies all of space. Not only that, it has a structure and stiffness to it. When a particle interacts with the Higgs field, the field attempts to hold onto the particle, to slow it down and resist its movement. Try to push a marble through molasses and you get the idea. Even if particles are inherently massless, when they interact with the Higgs field, they take on some of the Higgs field's rigidity. This is the phenomenon that we describe as mass. Particles that interact a lot with the Higgs field have a high mass. Particles that interact just a little with the Higgs field have a low mass. Some particles don't interact at all with the Higgs field. These are the massless particles that zip around at the speed of light, like the photon.

All of this was just a theory back in 1964. Experimental evidence was needed in order to prove that Higgs was right. The most promising way to validate the theory was to find the Higgs boson. We knew from the math that the Higgs particle needed to have charge of 0, a spin of 0, and a very large mass of around 125 GeV/c^2. It would take a very high energy to produce free Higgs bosons. And they would be very unstable, so would delay into other particles very quickly. For the balance of the 20th century, there was no way to find and verify the existence of the Higgs particle.

However, the Large Hadron Collider was completed in 2008, and for the first time, we had an instrument powerful enough to possibly produce and detect the Higgs boson.

This is the Large Hadron Collider, the world's largest particle accelerator. It is about 11 miles in diameter and straddles the border of France and Switzerland.

This is a look inside the Large Hadron Collider.

The search for the Higgs became one of the highest priorities for the LHC once it was operational. And in 2013, after months of verifying and cross checking the data, the LHC team announced they had found the

Higgs particle. It was the culmination of a 50 year search for experimental verification – a search which had given the Higgs particle the popular media name "the God particle".

With the discovery of the Higgs boson, Peter Higgs was finally awarded the 2013 Nobel Prize in Physics for his work.

The Higgs field gives space some shape and rigidity. As much as anything else we've talked about, it is probably the thing that makes space a something.

Now, for completeness, we should probably add a couple of subtle points here.

First, since $E = mc^2$ according to Einstein, where m is mass, giving anything mass requires supplying energy. The Higgs field supplies that energy. This implies that empty space is filled with a background non-zero energy level.

Second, the mass that particles acquire from the Higgs field is what is called rest mass. Ordinary matter actually gets most of its mass from the additional binding energy that holds the nucleus together. This binding energy comes from the strong nuclear force.

Now let's say one more thing about the Higgs field. You might at this point be wondering if it is the Higgs field that is being warped when we speak of gravity being the warpage of space by matter. The answer is no. Although the Higgs field conveys mass to particles, that is all it does. Gravity is still something else entirely, at least at this point in time. In other words, if there is an inherent connection, it is still shrouded in the mystery that we have yet to solve: how do we properly integrate general relativity and quantum theory?

Chapter 19 – Putting It All Together

So here we are at Chapter 19. We have covered a lot of subjects, and shed some light on the nature of space. But we are still struggling to answer our original question – the one posed by the title of this book: What is Empty Space?

The answer today is incomplete, because the question stands at the forefront of what we still don't know in physics today. But we can now summarize what we do know and at least partially answer the question.

We now know that space is a something. It's a tangible something. It's not static like Newton thought, but dynamic and changing, just as Einstein described. It can stretch and warp. It can twist and vibrate. It's a something, even though it looks like nothing!

Like water, space is a something we can see through. But also like water, it can be moved. Waves can be made in it. Those waves travel through it at the speed of light. Empty space has various properties we can examine and measure.

To go further, we must return to the subject we discussed in Chapter 5 – quantum field theory. Space is all about fields and the energy in those fields. The universe is basically made up of fields which fill all of space in varying degrees. Particles, including all matter, are simply perturbations (or disturbances) in these fields. When a particle acquires mass, it absorbs energy, an amount dictated by $E = mc^2$. But it is still a perturbation in its field. So an electron, for example, although it is clearly a particle with mass, is actually a disturbance or ripple in the electron field.

It is the distribution of these fields in space-time that makes our Universe what it is. The Higgs field is unique among all these fields in that it is distributed evenly across all of space. It's natural, lowest energy state is to exist everywhere uniformly. In this sense, it gives a certain tangible character to empty space itself.

In contrast, all the other fields tend toward dissipating – that is to spread out and tend toward a natural, lowest energy state of zero or non-existence. The gravitational field, for example, is strong around matter, but gradually dissipates as its distance from that matter increases, specifically following the inverse square law ($1/r^2$).

Empty space is a soup of quantum fields. The interaction of these quantum fields is what makes up our universe. The word "quantum" is the key here. Everything is quantized, most likely including space and time! That means only certain energy states are allowed. And therefore only certain field interactions are allowed. This in turn is what gives rise to the famous Standard Model in quantum field theory and string theory. It's what gives rise to all the particles and matter that make up our tangible universe.

If all of this sounds very theoretical and totally detached from the reality that we experience, you are not alone. It feels that way to many professional physicists. It may help to remember that solid matter, even when viewed in the absence of quantum theory, is mostly empty space. It only feels solid because of the repulsive force of the electric field generated by the electrons surrounding the nucleus of atoms.

So that's what we know about empty space. There are several things we don't know as well.

Dark energy, the force that is accelerating the expansion of the universe, is one of these things. Like the Higgs field, it apparently fills all of space. But is it a 5th natural force, like the strong force or gravity, or something else? We just don't know yet how it fits into nature and the Standard Model.

And what is the true nature of gravity? In spite of all of Einstein's brilliant insights into the nature of gravity, it's still full of mysteries one hundred years later. What is the true nature of this process by which matter warps space? And what is the quantum nature of gravity?

Then there are also questions that arise in the study of black holes. What is the nature of space near or inside the event horizon of a black hole? And why does empty space generate photons in the presence of the acceleration of mass?

Answering all these questions requires two things. The first is a lot more data.   It takes a lot of people and money to get that data. And the other thing that's needed is one great idea that makes everything clear. That perhaps only requires one smart mind and a new great insight!

The data required is coming. A group in Europe called the Euclid Consortium is working with the European Space Agency on a space mission that includes multiple instruments to map the universe and the effects of dark matter and dark energy with much greater accuracy than anything we have today. The Euclid space mission is scheduled to launch in 2020.

Another space instrument called the Wide-Field Infrared Survey Telescope (WFIRST) is a NASA observatory designed to perform wide-field imaging and surveys of the near infrared sky. Its planned launch is around 2025. Like Euclid, it promises a lot of new data on dark energy and the distribution of mass in the universe.

And back here on Earth, the Dark Energy Spectroscopic Instrument (DESI) will measure the effect of dark energy on the expansion of the universe from the Kitt Peak Observatory in Arizona. It will produce a highly detailed three dimensional map of the nearby universe out to 10 billion light-years, starting in 2018.

As a result of all this activity, a lot more data will be available in the next 20 years. But will that translate into a better understanding of dark energy, gravity, and empty space? Perhaps, but it's also possible we will still be waiting for another Einstein and some great new insight!

We still have a lot to learn about dark energy, gravity and empty space! Hopefully, a better understanding of empty space will help us get to a theory of everything. And hopefully, a good theory of everything will lead us to a full understanding of the nature of empty space.

Chapter 20 – Mathematics and Physics

We have tried to tackle some fairly difficult topics in this book without resorting to the extremely complicated mathematics which typically dominates the field of theoretical physics.   We have done that for three reasons.

First, this book is written for someone who wants to learn a little about physics, but isn't a physicist, and doesn't want to be subjected to intense mathematics.

Second, although the math frequently describes what is really going better than any non-mathematical description, we are trying to describe how this stuff works as something we can visualize, rather than just a mathematical abstraction.

And third, I couldn't explain a lot of this math even if I wanted to. I got far enough to take an elementary course in quantum mechanics, but the math behind quantum field theory and string theory is truly mind-boggling!

Behind every topic in this book, the knowledge we have today was coaxed and teased out of nature through mathematics. So it is fair to ask why mathematics is so important here. Why does math provide the best description of nature? Why is physics so dependent on mathematics?

If you go back to relatively simple stuff, the answer is fairly straight forward.   Isaac Newton described his laws of motion in mathematical terms because the things he was describing were defined mathematically to begin with. For example, velocity is defined as distance per unit time. Acceleration is change in velocity per unit time.   So, simply as a result of these definitions:

Distance = Velocity x Time

Or, adding acceleration to mix:

Distance = Velocity x Time + ½ x Acceleration x Time^2

Even the great work of Einstein is classical deterministic mathematics – relatively straight forward.   In fact, Einstein was a genius in his ability to visualize and conceptualize some very unintuitive concepts. He probably wasn't a mathematics genius. In fact, he was quite willing to get help on the math part from some of his contemporaries.

But since Newton's and Einstein's time, with the introduction of quantum theory and other non-deterministic physics, the mathematics involved has become more and more abstract and incredibly complex.   More and more, there is no obvious reason why a particular mathematical abstraction is the actual description of reality.

$$\mathcal{L}_{GWS} = \sum_f (\bar{\Psi}_f (i\gamma^\mu \partial\mu - m_f)\Psi_f - eQ_f \bar{\Psi}_f \gamma^\mu \Psi_f A_\mu) +$$

$$+ \frac{g}{\sqrt{2}} \sum_i (\bar{a}_L^i \gamma^\mu b_L^i W_\mu^+ + \bar{b}_L^i \gamma^\mu a_L^i W_\mu^-) + \frac{g}{2c_w} \sum_f \bar{\Psi}_f \gamma^\mu (I_f^3 - 2s_w^2 Q_f - I_f^3 \gamma_5) \Psi_f Z_\mu +$$

$$- \frac{1}{4} |\partial_\mu A_\nu - \partial_\nu A_\mu - ie(W_\mu^- W_\nu^+ - W_\mu^+ W_\nu^-)|^2 - \frac{1}{2} |\partial_\mu W_\nu^+ - \partial_\nu W_\mu^+ +$$

$$-ie(W_\mu^+ A_\nu - W_\nu^+ A_\mu) + ig' c_w (W_\mu^+ Z_\nu - W_\nu^+ Z_\mu)|^2 +$$

$$-\frac{1}{4} |\partial_\mu Z_\nu - \partial_\nu Z_\mu + ig' c_w (W_\mu^- W_\nu^+ - W_\mu^+ W_\nu^-)|^2 +$$

$$- \frac{1}{2} M_\eta^2 \eta^2 - \frac{gM_\eta^2}{8M_W} \eta^3 - \frac{g'^2 M_\eta^2}{32 M_W} \eta^4 + |M_W W_\mu^+ + \frac{g}{2} \eta W_\mu^+|^2 +$$

$$+ \frac{1}{2} |\partial_\mu \eta + iM_Z Z_\mu + \frac{ig}{2c_w} \eta Z_\mu|^2 - \sum_f \frac{g}{2} \frac{m_f}{M_W} \bar{\Psi}_f \Psi_f \eta$$

$$\sum_{m=1}^{M} \int_t \Big\{\Big\{ -\int_0^{2\pi} [\mathcal{L}_m(N_{xm} - \bar{N}_{xm}) a_m \delta u_m + \mathcal{L}_m(Q_{vm} - \bar{Q}_{vm}) a_m \delta v_m$$

$$-\mathcal{L}_m(Q_{wm} - \bar{Q}_{wm}) a_m \delta w_m + \mathcal{L}_m(M_{xm} - \bar{M}_{xm}) a_m \delta\psi_m]_0^l \, d\varphi$$

$$+ \int_0^l \int_0^{2\pi} \Big\{ \Big[\frac{\partial \mathcal{L}_m N_{xm}}{\partial x} + \Big(G\frac{t_m}{a_m^2}(1 + k_m^* I_m) \frac{\partial^2}{\partial\varphi^2} - \rho t_m \frac{\partial^2}{\partial t^2} \mathcal{L}_m \Big) u_m$$

$$+ G\frac{t_m}{a_m} \frac{\partial^2 v_m}{\partial\varphi \partial x} + \Big(G\frac{t_m}{a_m} k_m^* I_m \frac{\partial^2}{\partial\varphi^2} + \rho a_m t_m k_m \frac{\partial^2}{\partial t^2} \mathcal{L}_m \Big) \psi_m + \mathcal{L}_m P_{xm} \Big] a_m \delta u_m$$

$$+ \Big[\frac{\partial \mathcal{L}_m Q_{vm}}{\partial x} + \frac{vE}{1-v^2} \frac{t_m}{a_m} \frac{\partial^2 u_m}{\partial\varphi \partial x} + \Big(\frac{E}{1-v^2} \frac{t_m}{a_m^2} \frac{\partial^2}{\partial\varphi^2} - \rho t_m (1 + 3k_m) \frac{\partial^2}{\partial t^2} \mathcal{L}_m \Big) v_m$$

$$+ \Big(\frac{E}{1-v^2} \frac{t_m}{a_m^2} \frac{\partial}{\partial\varphi} + \rho t_m 2k_m \frac{\partial^3}{\partial\varphi \partial t^2} \mathcal{L}_m \Big) w_m - \frac{vEt}{1-v^2} k_m^* \frac{\partial^2 \psi_m}{\partial\varphi \partial x} + \mathcal{L}_m P_{\varphi m} \Big] a_m \delta v_m$$

$$- \Big[\frac{\partial \mathcal{L}_m Q_{wm}}{\partial x} + \frac{vE}{1-v^2} \frac{t_m}{a_m} \frac{\partial u_m}{\partial x} + \Big(\frac{E}{1-v^2} \frac{t_m}{a_m^2} \frac{\partial}{\partial\varphi} + \rho t_m 2k_m \frac{\partial^3}{\partial\varphi \partial t^2} \mathcal{L}_m \Big) v_m$$

$$+ \Big(\frac{E}{1-v^2} \frac{t_m}{a_m^2} \Big[1 + k_m^* I_m \Big(1 + \frac{\partial^2}{\partial\varphi^2} \Big)^2 \Big] + \rho t_m \frac{\partial^2}{\partial t^2} \mathcal{L}_m - \rho t_m k_m \frac{\partial^4}{\partial\varphi^2 \partial t^2} \mathcal{L}_m \Big) w_m$$

$$+ \frac{vEt}{1-v^2} k_m^* \frac{\partial^3 \psi_m}{\partial\varphi^2 \partial x} + \mathcal{L}_m \Big(P_{rm} + P_{rm}^{(vdW)} \Big) + \frac{1}{a_m} \frac{\partial m_{\varphi m}}{\partial\varphi} \Big] a_m \delta w_m \Big\} d\varphi \, dx \Big\} dt = 0.$$

Typical equations in String Theory

119

So it seems fair to ask why physics is so dependent on math. One straight-forward answer is simply that it works.

All the sciences including physics rely on experimental evidence and data. We know a theory is good when we can predict results from it, and then go on to verify those results experimentally.

Making one prediction and seeing that experimental results match that prediction is not good enough though. But making a whole variety of predictions based on a theory, and having them all turn out to be accurate under all kinds of conditions, makes the theory look pretty good.

Physicists express their ideas in mathematics, which can be tested experimentally. The ones that can pass rigorous experimental testing gradually get accepted and become the building blocks for even more detailed theories. Those that don't hold up under rigorous scrutiny simply get abandoned.

So gradually our knowledge pushes forward. But considering that our underlying faith in physics is that nature must be following some elegant simple rules, it is strange how this process has taken us to such abstract mathematics – certainly the math is very elegant, but far, far from being simple!

Most physicists just take it for granted that their chosen field is dominated by math. But some find it mysterious that mathematics takes such a pivotal role in understanding nature. Paul Dirac, the father of quantum field theory who we talked about in Chapter 5, lectured on the role of math in physics. He said, "There is no logical reason why using mathematical reasoning to learn about nature should be possible at all, but one has found in practice that it does work and meets with reasonable success. This must be ascribed to some mathematical quality in nature, a quality which the casual observer of nature would not suspect, but which nevertheless plays an important role in nature's scheme."

Einstein also wondered why nature's ways can be so neatly explained through mathematics when he said," "God is subtle, but not malicious.", and "Nature conceals her mystery by means of her essential grandeur, not by her cunning".

Physics in practice is all about mathematics. But the math that works and predicts experimental results is sometimes so abstract as to bear little resemblance to the reality it is trying to describe.

Final Thoughts

We now know a lot about empty space. We know it's teaming with various particles and fields, most of which we don't normally interact with. Gravitational fields are an exception, as we clearly interact with gravity.

The Higgs field is a special field that fills all of space. It gives space a structure and a rigidity that gives matter mass and inertia. Dark energy also fills all of space. It exerts a pressure on the universe as a whole, which is accelerating its expansion. But it is not yet fully understood.

Einstein told us that space can be warped, stretched, and even broken, as happens at the event horizon of a black hole. Space can spontaneously create particles. And we know space itself is probably quantized!

But our understanding of empty space is still very incomplete. The nature of empty space is at their core of many of the great mysteries in physics today.

In spite of all the great discoveries in physics over the last 300 years, we are still like a deep ocean fish. She lives in water. It's everywhere around her. She has never experienced anything else! It's her version of empty space. She can see through it. She can freely move through it. So to her, it's simply empty space.

Perhaps if we could see our Universe from a completely distant and different perspective, we might view empty space the way we view water today. It would be a something very tangible. It would have very specific, very well understood properties.

We can't see space like that yet. We are sort of like that fish. But we are also different than the fish in one respect. She will never perceive the existence of water. She will never try to contemplate its nature.

But we now do perceive the existence of empty space. It is a real, tangible something. We are contemplating its nature. We can experiment and eventually learn exactly what it is. We are making progress. We are getting closer.

- ∞ -

Index

About the Author

Doug Domke is not a physicist by profession. He does have a Bachelor of Science degree in Physics, but spent his career in the semiconductor / computer-chip industry. He is currently retired and lives in Phoenix, Arizona.

In addition to "What is Empty Space?", Doug is also the author of:

"A Small Book on Physics" (2012)

Each chapter is only a few pages, but gives you an overview of an entire subject. One chapter tells you all about "black holes". Another tells you what "quantum mechanics" is. You will also learn about the more classical areas of physics. What is "thermodynamics" or the "physics of sound"? You'll learn what physicists still don't understand, where they are in their quest, and what's likely to happen next! Other topics include: The Big Bang, Gravity before and after Einstein, String Theory, The size and age of the Universe, Electromagnetic Waves, Solid State Physics, Lasers and Holograms, as well as Sub-atomic Particles.

"What's Inside the Black Box?" (2014)

Technology is all around us today, and for most of us, the inner workings are incomprehensible. What you might not know is that most of the inner workings are incomprehensible to even the people who designed them. That's because modern technology is the result of collective intelligence, where one person may make a significant contribution, but doesn't know and doesn't need to know how the complete system works. We get great technology by compartmentalizing complex technological challenges. One guy figures out how a certain input generates a specific output. Once that's done, everyone can use it. The inner workings can remain unknown to everyone else. It's what's known as a "black box". In this book we will examine a number of items you use every day, like your microwave oven, digital camera or automatic transmission. We'll take them apart and see how they work. Then you will know "What's Inside the Black Box"!

www.ingramcontent.com/pod-product-compliance
Lightning Source LLC
Chambersburg PA
CBHW022007170526
45157CB00003B/1178